The Open University

ence: A Second Level Course

BIOCHEMISTRY

3 Cell Energetics, the Mitochondrion and the Chloroplast
4 Metabolic Pathways

Prepared by an Open University Course Team

THE OPEN UNIVERSITY PRESS

THE BIOCHEMISTRY COURSE TEAM

Steven Rose (*Chairman and General Editor*)
Norman Cohen
Jeff Haywood
Brian Tiplady
Eve Braley-Smith (*Editor*)
Robin Harding (*Course Assistant*)
Bob Cordell (*Staff Tutor*)
Vic Finlayson (*Staff Tutor*)
Roger Jones (*BBC*)
Jim Stevenson (*BBC*)

The Open University Press
Walton Hall, Bletchley, Buckinghamshire

First published 1972

Designed by The Media Development Group of the Open University.

Printed in Great Britain by
Staples Printing Group
at St Albans

SBN 335 02051 8

This text is one in a series of units that make up the correspondence element of an Open University Second Level course. The complete list of units in the course is given at the end of this text.

For general availability of supporting material referred to in this book, please write to the Director of Marketing, The Open University, Walton Hall, Bletchley, Buckinghamshire.

Further information on Open University courses may be obtained from the Admissions Office, The Open University, P.O. Box 48, Bletchley, Buckinghamshire.

1.1

UNIT 3

CELL ENERGETICS, THE MITOCHONDRION AND THE CHLOROPLAST

Contents

Table A List of Scientific Terms, Concepts and Principles Used in Unit 3

Taken as prerequisites			Introduced in this Unit			
1 Assumed from general knowledge	2 Introduced in a previous unit	Unit No.	3 Developed in this Unit or in its set book	Page No.	4 Developed in a later Unit	Unit No.
		S100[1]				
		or				
		C of L[2]		**S2–1**		
	activation energy	11	coenzyme A (CoA.SH)	17	citric acid cycle	4
	adenosine triphosphate	15	energy transduction	13	glycolytic pathway	4
	(ATP)		exergonic, endergonic		$NAD/NADPH_2$	
	adenosine diphosphate	15	reactions		ratio	4
	(ADP)		FAD, flavoprotein	11–12	pentose phosphate	
	aerobic	15	free energy, (G, ΔG, $\Delta G^{\ominus}$)	22	shunt	4
	algae	18, 34	glyceraldehyde-3-	10	mitochondrial/	
	amino acid	10, 14	phosphate		cytoplasmic	4
	anabolism	15	dehydrogenase		interactions	
	anaerobic	15	group transfer molecules	20	control processes	
	catabolism	15	CTP, GTP, UTP	15		4
	chemotrophic	21	lipoic acid (LSH)	15		
	chloroplast	16	$NADH_2$	20		
	citric acid cycle	15	dehydrogenase			
	(Krebs cycle)		Slater criteria for	26		
	creatine, creatine	15	intermediates			
	phosphate (CP)		steady state	26		
	cytochrome, cytochrome	15	submitochondrial	13		
	oxidase		complexes			
	cytoplasm	14	substrate level	27		
	DNA	14	phosphorylation			
	electron transport	15	thiamine pyrophosphate	20		
	endoplasmic reticulum	14	(TPP)			
	endothermic	11	thioester	20		
	entropy (S, ΔS)	11		16		
	equilibrium constant	11		**CS &**		
	(K)		ATPase	**F**[3]		
	eucaryote	17	Calvin cycle	326		
	exothermic	11	chemiosmotic theory	340		
	fatty acid	14	chlorophyll	322		
	glucose oxidation	15	chloroplast structure	332		
	glycolysis	15	coenzyme Q	336		
	heat of reaction (ΔH)	11	(ubiquinone)	317		
	hydrogen ion, H^+	12	coupled oxidation			
	hydrolysis	12	cristae	318		
	hydroxyl ion, OH^-	12	crossover sites	312		
	lipid	14	cytochrome oxidase	320		
	lipoprotein	14	cytochromes a, b, c, c_1	317		
	lysosome	14, 16	dark reaction	317		
	metabolism	15	electron transport	338		
	methylene blue	15	particles	323		
	mitochondria	14, 15	(submitochondrial			
	NAD, NADP	15	particles)			
	nucleus	14				

[1] *The Open University (1970) S100* Science: A Foundation Course, *The Open University Press.*

[2] *Steven Rose (1971)* The Chemistry of Life, *Penguin Books: Open University Edition.*

[3] *A. G. Loewy and P. Siekevitz (1969)*, Cell Structure and Function, *2nd ed., Holt, Rinehart & Winston.*

Taken as prerequisites			Introduced in this Unit			
1 Assumed from general knowledge	2 Introduced in a previous unit	Unit No.	3 Developed in this Unit or in its set book	Page No.	4 Developed in a later Unit	Unit No.
	oxaloacetic acid	15	grana	336		
	oxidative phosphorylation	15	'high energy intermediate' theory	321		
	α-oxoglutaric acid	15	Hill reaction	335		
	phage	17	matrix	314		
	phosphate, inorganic (Pi)	15	mitochondrial DNA	342		
	pyruvic acid	15	mitochondrial ion transport	323		
	quantum	29	mitochondrial membranes	313		
	ribosome	14, 17	mitochondrial protein synthesis	344		
	stoicheiometry	9	mitochondrial structure	313		
		S100	mitochondrial swelling	327		
	symbiosis	20	mitochondrion as a multi-functional organelle	342		
	urea	21	phosphorylating electron transport particle	323		
	bond cleavage	11	photosynthetic phosphorylation	333		
	conjugation	10	plastid	72		
	electron acceptor	8	plastoquinone	333		
	electron donor	8	P/O ratio	318		
	esterify	10	quantasome	75		
		S24–[4]	stopped-flow and steady-state methods	320		
	nucleophilic substitution	5	succinic dehydrogenase	317		
	resonance	6				
		S2–1				
	coenzyme	2				
	enzyme	2				
	inhibitor	2				
	multienzyme complex	1, 2				
	sulphydryl	1				
	thiamine	2				
	tonometer	TV 3				
	vitamin	2				

[4] *The Open University (1971) S24–* An Introduction to the Chemistry of Carbon Compounds, *The Open University Press.*

Objectives

When you have completed this Unit, you should be able to:

1 Demonstrate knowledge of items in Table A by being able to define them in your own words, or recognize the best definition or description of them. (*SAQ*s 3, 7, 17, 18, pp. 23, 31.)

2 When presented with experiments concerned with cell respiration and energetics (not necessarily discussed in the Unit), distinguish between warranted, unwarranted and contradicted conclusions drawn from them. (*SAQ* 1, p. 14.)

3 Understand the limitations of, and discuss the advantages and disadvantages of, the application of thermodynamic concepts to cellular metabolism. (*SAQ*s 2, 4, 7, pp. 16, 23.)

4 Distinguish, among a given series of molecular structures, which are likely to be able to serve, and which are likely to be unable to serve, as group transfer molecules. (*SAQ*s 2, 4, 7, pp. 16, 23.)

5 Assess, and justify the assessment by reference to particular systems, the relative contribution of different metabolic pathways and cell organelles to the overall ATP-generating capacity of the cell. (*SAQ* 20, p. 33.)

6 Distinguish, with examples, at least four different types of ATP-generating reactions. (*SAQ*s 3, 5, 6, 8, 9, pp. 23, 24.)

7 Describe, and assess the limitations and merits of and possible conflicts between at least three different experimental approaches to the study of mitochondrial functions. (*SAQ*s 12, 19, pp. 26, 32.)

8 Name and ascribe a place in the carrier sequence to at least six components of the electron transport chain. (*SAQ*s 13, 14, 15, 16, 17, pp. 27, 28, 31.)

9 Draw a diagram or otherwise indicate knowledge of the ultrastructure of the mitochondrion and the possible localization of enzymic and electron transport function within it. (*SAQ*s 9, 10, 11, 12, 14, pp. 24, 26, 28.)

10 Distinguish between the effects of inhibiting the electron transport chain and either uncoupling or inhibiting the mechanism of phosphorylation. (*SAQ* 21, p. 33.)

11 Indicate the presumed sites of coupling of ATP synthesis to the electron transport chain. (*SAQ* 22, p. 33.)

12 Suggest the possible physiological significance of the ionic properties of the intact mitochondrion and the mitochondrial/cytoplasmic interactions.

13 Provide up to four examples justifying the concept of the mitochondrion as a multifunctional organelle, and speculate as to the implications for the evolutionary origin of the mitochondrion. (*SAQ* 23, p. 34.)

14 Assess the similarities and differences between mitochondrial and chloroplast functions. (*SAQ* 14, p. 28.)

15 Describe in outline, using a flow chart, the chief events involved in photosynthesis, including light and dark reactions and the Calvin Cycle.

Study Guide and Preliminary Reading

Unit 3, concerned with the question of ATP, the role of the mitochondria and photosynthesis, derives most directly from the central section of S100, Unit 15; particularly Sections 15.4, 15.6, 15.7 and 15.8 and its set book, *The Chemistry of Life*. Before embarking on Unit 3, you should revise this material thoroughly, omitting the structured exercise and Home Experiment, of course, but including the black-page reading from *Chemistry of Life*. In order to make this revision, first check yourself by doing *SAQ*s 8–12 of S100, Unit 15 (note the error in the printed answers to *SAQ* 16 in the first edition of the course material). If you can correctly answer them all, then you obviously don't need to revise further. If you answer some correctly but not others, you need to revise only the sections you are shaky on.

If you feel you need to make an extensive revision, you may prefer as an alternative procedure to go directly to *Chemistry of Life* and read through Chapters 6 and 7, pp. 110–49, and pp. 191–6, rather than to work through the Unit text. But re-check yourself on the *SAQ*s afterwards! If you did not read the black-page material from *Chemistry of Life* last year, this will be new to you. If you are in doubt as to terms you are expected to be able to recall in order to begin this Unit, check with Table A.

We also make use in this Unit of concepts of reaction mechanisms and thermodynamics, dealt with in S100, Units 11 and 12, Sections 11.1.3, 11.1.4, 11.2.3 and 12.4.1. Only when you have satisfied yourself that you understand this material should you proceed to work on the new material of this Unit. In doing so, you should note that the prescribed reading for the Unit includes the whole of Chapter 13 of *Cell Structure and Function*, pp. 310–44, about 12 000 words in all, of which some is essentially recapitulation of material from *Chemistry of Life*, and some (the photosynthesis section (about 3 800 words)) may be partly revision for some of you, if you have done S22–, the Comparative Physiology Course.

Because our approach in the Unit is not quite the same as that of *Cell Structure and Function*, we do not refer out to consecutive sections of that text, but to sections in a non-sequential order. You may therefore prefer to approach work on this Unit by first reading the whole of Chapter 12 of *Cell Structure and Function* reasonably rapidly, without trying to 'learn' the material. Then start the Unit text, referring back to *Cell Structure and Function* at the prescribed places, and studying the material more thoroughly.

The following table indicates where in the Unit there is prescribed reading:

Section	*Set Material*
3.1	—
3.2	*Cell Structure and Function*, pp. 15–16 (1 000 words)
3.3	—
3.4	—
3.5	—
3.6	*Cell Structure and Function*, pp. 310–14 (1 800 words)
3.7	*Cell Structure and Function*, pp. 316–18 and 323–6 (1 800 words)
3.8	*Cell Structure and Function*, pp. 318–23 (2 500 words)
3.9	*Cell Structure and Function*, pp. 326–30 and 342–4 (2 000 words)
3.10	*Cell Structure and Function*, pp. 331–42 (3 800 words)

Note that Section 3.10 is concerned with photosynthesis. It is this Section which will be familiar, though from a different viewpoint, to students from S 22–.* Although it is important from the point of view of completing some of the Unit Objectives, it is self-contained in that subsequent material in Units 4–6 of the Course does not depend on it. If you are running short of time, you are advised to omit Section 3.10 of the Unit.

* *The Open University (1972) S22–*, Comparative Physiology, *The Open University Press.*

3.1 Introduction

In the last two Units, you have learned something about the structure of proteins and the working of enzymes; that is, about one of the key concepts of this Course, the structure/function relationship. In this Unit, we turn to another of the key concepts of biochemistry, albeit a more vexed and disputed one, that of cell energetics. In S100, Units 14–16, we suggested that the history of biochemistry was a history of asking more and more complex questions about cell processes. One of the questions we singled out as of major significance was that of metabolism, of the routes and mechanisms by which one chemical within the cell was converted to another. In our discussion of metabolism, we pointed out (S100[1])* that in the cell one was dealing with reactions many of which were strongly endothermic (S100[2]). One of the problems with which biochemists have been concerned, therefoie, is that of how endothermic reactions are 'driven' in the cell. The picture we presented in S100, Unit 15, was of endothermic reactions being driven by coupling them to exothermic (S100[2]) reactions. This chemical coupling was achieved, we said, by way of a coenzyme of almost universal biochemical significance, the substance adenosine triphosphate, ATP (S100[1]).

* *The list of references to S100 is on p. 39.*

The analysis of biochemical processes within the cell in terms of energetics and energy transfer is one which has been of great concern to biochemists; in fact, it dominated the history and thinking of biochemists through the 1940s and 1950s and into the 1960s. Many if not all of cell functions were analysed in terms of the *energy traffic* involved. This type of analysis still dominates many of the biochemical textbooks. Although relatively less dominant in your set text, *Cell Structure and Function*, you will find this type of approach in such recommended background reading books as Conn and Stumpf, Mahler and Cordes, Baldwin, and Lehninger.

Over recent years, however, there has been a significant shift in biochemical thinking, the type of central concepts which the biochemist uses, and therefore the questions he asks of his material and the types of experiment he designs in order to try to answer them. These new questions are formulated in terms of ideas such as those of *regulation* and *control*; what determines the rates of reaction sequences and how the cell regulates the amounts of particular substances produced at any one time. The introduction to these themes was provided in S100, Unit 16, and they are the principal topics of this Course.

At the same time the somewhat cavalier approach of the biochemist to chemical thermodynamics has come in for severe criticism from some chemical thermodynamicists on the grounds that it is theoretically unsound to attempt to analyse cell reactions in thermodynamic terms at all. We hinted at this conflict of approach in S100[1] in our commentary on the treatment of ATP and the 'squiggle' notation (~) in *Chemistry of Life*.

In this Unit we shall begin by presenting the 'classical' biochemical view of cell energetics, and then discuss in outline the nature of the attack upon this view. By doing this, we will make it possible for you to follow the treatment of energetics given in many standard biochemical textbooks (Section 3.2).

With a better understanding of the theoretical background we can turn to the formulation of an acceptable statement of the role of group transfer molecules like ATP and some other phosphates in the coupling of reactions within the cell (Section 3.3) and discuss the cellular roles of ATP (Section 3.4). We revise a number of reaction mechanisms whereby ATP is generated non-oxidatively (Section 3.5), before turning to a more detailed discussion of the principal ATP generating systems within the cell, the process of oxidative phosphorylation (S100[2]).

In Section 3.6 we consider the structure of the mitochondrion in relation to its function. Within the mitochondrion, we turn first to the electron transport chain, considering the variety of techniques available for its study and the conclusions which can be drawn from them. Models of oxidative phosphorylation mechanisms are dealt with in Section 3.8. In Section 3.9, we turn to the relationship between structure and function in the mitochondrion in the control of ATP generating and other energy-providing reactions; the physiological roles of the mitochondrion. Finally, in Section 3.10, we consider the related structure, the chloroplast, and the mechanism of photosynthesis.

3.2 The Thermodynamic Approach

The problem which concerns us here is that of *what causes a reaction to take place*. Those reactions which can occur, in any given set of circumstances, are said to be thermodynamically feasible, those which cannot are thermodynamically unfeasible. The catalysis of a reaction by an enzyme can never make a thermodynamically unfeasible reaction into a feasible one, it can merely speed up the rate at which the feasible reaction occurs. In S100, we rather slid over the question of what makes a reaction thermodynamically feasible or unfeasible, but it is now necessary to treat this slightly more respectably, although still not very rigorously.

feasible and unfeasible reactions

First, remind yourselves of the meaning of the terms *entropy*, *S*, *enthalpy*, *H*, and the equilibrium constant of a reaction, *K*, introduced in S100. If you do not recall these terms, check back first to S100[5] for the definition of the term ΔS, the change in entropy, and then to S100[6] for the terms enthalpy, exothermic and endothermic reactions and the definition of the term ΔH, and to S100[7] for *K*. In S100, no attempt was made to relate these terms, but we must now try to do this.

entropy
enthalpy

Classical thermodynamics deals with the equilibrium of a reaction. For example in the schematic reaction

$$A+B \rightleftharpoons C+D$$

the equilibrium constant is derived from the concentrations of the products and reactants in an equilibrium mixture under defined conditions by the equation

equilibrium

$$K \rightleftharpoons \frac{[C]\,[D]}{[A]\,[B]}$$

Another function is more commonly used to define the equilibrium position, called the *standard free energy change* of the reaction, $\Delta G^{\ominus}$* expressed by the relationship

$$\Delta G^{\ominus} = -RT\ \ln K.$$

where R is the gas constant (S100[8]), T is the temperature in degrees Kelvin above absolute zero (S100[9]) and ln*K* is a value derived from *K* (do not worry how at this stage, as it doesn't matter for what follows). This equation enables $\Delta G^{\ominus}$ to be measured; an example of this measurement is given in *Cell Structure and Function*, pp. 15–16, which you should now read.

standard free energy

Looking at the equation for $\Delta G^{\ominus}$ you should be able to see that if the reaction

$$A+B \rightleftharpoons C+D$$

has its equilibrium far to the right, it will have a *large negative* $\Delta G^{\ominus}$. If the equilibrium mixture contains equal amounts of all four compounds A, B, C and D, the $\Delta G^{\ominus}$ is zero, and if the equilibrium is to the left, the $\Delta G^{\ominus}$ will be positive.

There are two reasons for using $\Delta G^{\ominus}$ instead of an equilibrium constant. The first is that $\Delta G^{\ominus}$ is directly related to $\Delta H^{\ominus}$, the standard enthalpy change, and $\Delta S^{\ominus}$ the standard entropy change, by the equation

$$\Delta G^{\ominus} = \Delta H^{\ominus} - T\Delta S^{\ominus}**$$

$\Delta G^{\ominus}$ is a measure of the total work that can be done by a chemical reaction, and

* *In S25-,* Structure, Bonding and the Periodic Law, *The Open University Press (1972), this term is introduced under its more modern name, the Gibbs function. The Term* G *is sometimes referred to as* F *in older textbooks. This only adds to the confusion as* F *also has a meaning in relation to quite another thermodynamic function.*

** *The superscript* $^{\ominus}$ *represents a function measured under standard conditions of temperature, pressure, and composition of solution. The values of free energy changes, etc., measured under different conditions vary widely and these general functions are represented by* ΔG, ΔH *and* ΔS. *Provided all three functions are determined under identical conditions*

$$\Delta G = H - T\Delta S$$

but in this Unit only the standard functions will be considered.

this forms the basis of another method of measurement, especially when a reaction proceeds essentially to completion, i.e. when the equilibrium mixture contains no measurable amount of A and B.

The second reason for using $\Delta G^{\ominus}$, which is more important in this context, is that free energy changes are additive. Thus if a reaction proceeds in two steps (or can be represented as two steps) the $\Delta G^{\ominus}$ of the total reaction is the sum of the $\Delta G^{\ominus}$ values of the two partial reactions. So if three compounds are in equilibrium:

$$\mathrm{A} \rightleftharpoons \mathrm{B} \rightleftharpoons \mathrm{C}$$

and the $\Delta G^{\ominus}$ changes are as follows:

$$\mathrm{A} \rightarrow \mathrm{B},\ \Delta G^{\ominus} = -20\ \mathrm{kJ\ mol^{-1}}$$
$$\mathrm{B} \rightarrow \mathrm{C},\ \Delta G^{\ominus} = +4\ \mathrm{kJ\ mol^{-1}}$$

then the $\Delta G^{\ominus}$, of the reaction A→C is

$$(-20+4) = -16\ \mathrm{kJ\ mol^{-1}}.$$

Note that if the $\Delta G^{\ominus}$ for A→C is -16 kJ $\mathrm{mol^{-1}}$, the $\Delta G^{\ominus}$ for C→A is $+16$ kJ $\mathrm{mol^{-1}}$. So the reactions must be written the right way round before adding them up.

Note, too, that it is not necessary for A to go to C via B (e.g. the reactions may be:

$$\begin{array}{c} \mathrm{A} \rightleftharpoons \mathrm{B} \\ \nwarrow\!\searrow \ \swarrow\!\nearrow \\ \mathrm{C} \end{array}$$

for this to be valid. So long as the stoichiometry of the two reactions add up to a third reaction, the $\Delta G^{\ominus}$ values will also add up correctly. So these thermodynamic sums tell us nothing about the *mechanism* of a reaction.

Clearly, the $\Delta G^{\ominus}$ of any overall pathway that occurs in the cell must be negative. This does not apply to the individual steps, however. So long as the products are being continuously removed, a reaction with a distinctly unfavourable equilibrium can proceed at the same rate as the rest of the pathway. Within limits, the $\Delta G^{\ominus}$ of individual steps in a pathway is not important. This effect of concentration of reactants and products on the rate of a reaction is known as *mass action*: a build-up of reactants speeds up the overall rate of a reaction, a build-up of products slows it down.

mass action

Many of the reactions of catabolism in the cell have large negative $\Delta G^{\ominus}$ values while anabolic ones tend to have positive $\Delta G^{\ominus}$ values when written in their simplest terms. This may be misleading. For instance the anaerobic breakdown of glucose in mammalian muscle produces lactic acid. The reaction

$$\text{glucose} \rightarrow 2\ \text{lactic acid}$$

has a $\Delta G^{\ominus}$ of -189 kJ $\mathrm{mol^{-1}}$.

But this is not the reaction that occurs, which is

$$\text{glucose} + 2\ \mathrm{ADP} + 2\text{Ⓟ}^{-} + 2\mathrm{H^{+}} \rightarrow 2\ \text{lactic acid} + 2\ \mathrm{ATP} + 2\mathrm{H_2O}.$$

To find the $\Delta G^{\ominus}$ of this reaction, the $\Delta G^{\ominus}$ of the partial reactions involving ATP must be added

$$2\mathrm{ATP} + 2\mathrm{H_2O} \rightleftharpoons 2\mathrm{ADP} + 2\text{Ⓟ}^{-} + 2\mathrm{H^{+}};\ \Delta G^{\ominus} + 62\ \mathrm{kJ\ mol^{-1}}$$

Thus the overall change is:

$$(-189+62) = -127\ \mathrm{kJ\ mol^{-1}}.$$

Cellular reactions thus tend to be coupled, so that part reactions which would, by themselves, have large negative $\Delta G^{\ominus}$ values are linked to other part reactions which would, by themselves, have large positive $\Delta G^{\ominus}$ values. One set of reactions can be used to 'drive' the other.

In chemical terms, what provides this coupling? One of the observations made repeatedly in the study of a variety of exergonic reactions during the 1930s was that they seemed to require participation of the inorganic phosphate ion ($H_2PO_4^-$), generally represented $Ⓟ_i^-$. During the reaction, the substance adenosine diphosphate (ADP) was consumed and adenosine triphosphate (ATP) appeared.

ATP

$$ADP + Ⓟ_i^- + H^+ \rightleftharpoons ATP + H_2O.$$

During other reactions in the cell, generally endergonic synthetic ones such as glycogen synthesis, ATP tended to be broken down again, the reaction above running from right to left. These reactions were explored predominantly in relationship to muscle,* and the detailed study of heat production that occurs during muscular contraction.

In 1927 Eggleton and Eggleton described creatine phosphate and three years later Lohmann isolated ATP from rabbit muscle. By the early 1930s the linked reaction which reversibly converts phosphate and ADP into creatine and ATP (the so-called Lohmann reaction) was mapped out, and Lundsgaard was able to show the breakdown of creatine phosphate in muscle which had been poisoned with iodoacetic acid (S100[10]) and its resynthesis in the presence of oxygen in normal muscle. These reactions were all related by Lundsgaard in 1931 to a model for the energy-linked synthesis of phosphate bonds. Studies of oxidative phosphorylation, as opposed to anaerobic muscle twitches, led to the suggestion, by Kalckar, Cori and Lipmann, of the existence of some intermediate phosphate compound which was synthesized during oxidation and subsequently participated in the synthesis of creatine phosphate.

creatine phosphate and muscle

Finally, from all these observations, Lipmann, in 1941, drew the conclusion that ATP formed the 'energy currency' of the cell; it was the proximate source of $(-\Delta G)$ necessary to drive synthetic reactions or cell activities such as muscular contraction, and was itself synthesized during glucose oxidation; it was the link in the coupled reaction sequence. Cell reaction sequences could then be analysed in terms of whether they were 'ATP generating' or 'ATP requiring'. ATP was one example of a class of compounds of generally 'high-energy character'. But what was there special about the ATP molecule which fitted it for this unique role? It seemed obvious that the right thing to do was to examine the reaction

ATP as 'energy currency'

$$ATP + H_2O \rightleftharpoons ADP + Ⓟ_i^- + H^+$$

for its interesting features. When the $-\Delta G^{\ominus}$ change in this reaction was determined, it was observed to be high relative to the free energy change occurring in such other analogous reactions as those of the hydrolysis of adenosine monophosphate (AMP) and glucose-6-phosphate (G-6-P)

$$AMP + H_2O \rightleftharpoons A + Ⓟ_i^- + H^+$$
$$G\text{-}6\text{-}P + H_2O \rightleftharpoons G + Ⓟ_i^- + H^+$$

In the two latter reactions, $\Delta G^{\ominus}$ is of the order of -12.5 kJ mol^{-1}, while in the hydrolysis of ATP to ADP or ADP to AMP, $\Delta G^{\ominus}$ is of the order of 31 kJ mol^{-1}; actually, the early measurements of this figure gave values as high as -40 to -50 kJ mol^{-1}. There was thus, Lipmann argued, something unique about the two terminal phosphate Ⓟ bonds of ATP which distinguishes them from the phosphate bond in AMP or G-6-P. They could be regarded, he claimed, as 'high energy bonds' compared with the 'low-energy' phosphate bonds in AMP and G-6-P.

This formulation is in fact misleading in a number of ways. For example, in the 'high energy compound' formulation, there is the idea that high energy compounds form a distinct class from 'low energy compounds'. In fact, if the $\Delta G^{\ominus}$ of hydrolysis is measured for a wide range of esters and anhydrides, a continuous range of values is obtained. It should also be emphasized that $\Delta G^{\ominus}$ is not the property of a *molecule* but of a *reaction*.

'high energy compounds'

Nonetheless, the term 'high energy bond' stuck, despite such criticisms, and Lipmann introduced the squiggle notation ~ to distinguish between the two terminal and the first phosphate bond of ATP. Later the concept was modified to refer instead to ATP as a 'high energy molecule' the relatively high free energy

* *See SDT 286, Unit 4, for a discussion of muscular contraction. (The Open University (1971) SDT 286:* Biological Bases of Behaviour, *The Open University Press.)*

of hydrolysis being a property, of course, of all the molecular structures participating in the reaction, not just the terminal phosphate bond of ATP. The ~ notation was then used to refer to the bond simply to indicate which bond was broken or resynthesized during the 'energy-transferring' reactions. From this analysis, the analogy of the cell's 'energy-bank' into which the oxidation of glucose put payments of ~P, while synthetic reactions or muscular contraction drew ~P out, was derived.

The approach has led to a long series of studies of the $\Delta G^{\ominus}$ of individual reactions in the glycolytic (S100[11]) or glucose oxidation (S100[12]) sequence to see at which steps the $\Delta G^{\ominus}$ of the reaction is more negative than -31 kJ mol^{-1} and can thus be used to drive ATP synthesis. Many textbooks analyse individual reactions in order to determine whether or not they are thermodynamically feasible and if ATP can be synthesized during them.

But as has been suggested already, the whole approach of analysing cellular reactions in thermodynamic (as opposed to kinetic) terms has come under attack recently, especially by chemical thermodynamicists such as Banks, Vernon and Ross. In a controversial area, the following points in their analysis may be singled out. Firstly, they point out, the thermodynamic equations such as that for $\Delta G^{\ominus}$ given above are applicable only to equilibrium reactions in a closed system. The cell is an open system; metabolites are constantly entering it and leaving it. If it were a closed system it would be dead. The thermodynamics of open systems are very complex and only a partial approach to their resolution is possible. Because the cell is an open system, in which reactions are constantly proceeding, it is not at equilibrium. Rather, it is in a steady state. To see what this means, consider the difference between the reversible reaction at equilibrium

closed and open systems

$$A+B \rightleftharpoons C+D$$

which we used above to show how $\Delta G^{\ominus}$ for a reaction could be calculated, and the actual situation in a cell which is best expressed

$$\begin{array}{ccccccc} \rightleftharpoons a & & & & & & d \rightleftharpoons \\ \upharpoonleft\downharpoonright & & & & & & \upharpoonleft\downharpoonright \\ A & + & B & \rightleftharpoons & C & + & D \\ & & \upharpoonleft\downharpoonright & & \upharpoonleft\downharpoonright & & \\ & & \rightleftharpoons b & & c \rightleftharpoons & & \end{array}$$

A and B are constantly being generated from percursors (a) and (b), themselves in their turn generated from elsewhere in the system. Similarly C and D are being converted to products (c) and (d), which are in turn removed.

In the normal working conditions of the cell, the concentrations of A, B, C and D will probably be constant, even though there is a constant throughput of molecules through the reaction. It is like a bath with taps on and plug out, always filling, always emptying. If the flow is adjusted appropriately, then the level of the bath water remains constant though the individual H_2O molecules are always changing (hence the more recent biochemical interest in what regulates the flow). Not only are enzymic reactions in the cell not at equilibrium, but many cell reaction sequences depend on continual removal of product, for the equilibrium of the reaction lies far in the 'wrong' direction from the point of view of the reaction sequence of which it is a part. Constant removal of products is essential. Indeed, ATP levels in the cell are a good example of just such a steady state situation with constant removal and constant synthesis. ATP levels rarely change unless substantial demands are placed upon the system (see below, Section 3.4) and the ADP/ATP ratio remains almost constant. Once again, thermodynamic analysis cannot be applied to this situation.

ADP/ATP ratio

Again, the critics go on to point out, all the calculations of $\Delta G^{\ominus}$ on which the designation of ATP as a high energy molecule are based depend on the measurement of the hydrolysis reaction

$$ATP+H_2O \rightleftharpoons ADP+\text{Ⓟ}_i+H^+$$

a reaction which does not normally occur with the cell—if it does, the 'free energy' of ATP is regarded as 'wasted' or dissipated; the normal reactions in which ATP is involved are not this at all, as we shall see, but generally the phosphorylation of other molecules. Thus calculations based on the thermodynamics of the hydrolysis reaction are misleading.

Finally, calculations based on the $\Delta G^{\ominus}$ of a reaction studied in a test tube may have little bearing on the ΔG that actually occurs in the situation of the cell, because ΔG varies according to the concentrations of the reactants, the pH, the presence of ions and the state of the reactants; in solution, bound into an insoluble lipoprotein matrix, such as the mitochondrion, or whatever.

These telling criticisms should lead to a reconsideration of the way biochemists formulate the role of ATP. It may be preferable to abandon thermodynamic considerations, because they are of limited utility, and instead regard ATP, as we shall do in the remainder of this Unit, as the chief representative of a class of molecules which are obligatory coenzymes for a variety of enzyme reactions. In these reactions, molecules are primed by one of several types of phosphorylation reaction for which ATP is the phosphate donor molecule. These are generally reactions involved in synthetic sequences within the cell. ADP and AMP, generated by these phosphorylations, are themselves phosphorylated during other cellular reactions, notably those of glycolysis, glucose oxidation and fatty acid oxidation (S100[13]).

Thermodynamic considerations do not enter here, but the phosphorylations are seen as required from kinetic consideration of the enzyme reactions, notably those of the activation energy for the reaction (S100[14]), (see also Unit 2 of this Course). We need not go all the way with the thermodynamicists in their critique in order to provide a general statement for the role of ATP as follows:

If in the reaction $X{-}Y \rightleftharpoons X+Y$, $\Delta G^{\ominus}$ is negative, then the reverse reaction, $X+Y \rightleftharpoons X{-}Y$ will occur, in biological systems, only if one of the relevant molecules is first converted into an alternative form, say $X{-}R$, so that in the reaction: $X{-}R+Y \rightleftharpoons X{-}Y+R$, $\Delta G^{\ominus}$ is also negative.

group transfer substances

The group 'R' may be referred to as the priming group and is frequently the phosphoryl radical derived from ATP, thus:

$$X+ATP \rightleftharpoons X{-}Ⓟ+ADP$$

$$X{-}Ⓟ+Y+H_2O \rightleftharpoons X{-}Y+Ⓟ_i+H^+$$

$$\text{SUM} \quad X+Y+ATP+H_2O \rightleftharpoons X{-}Y+ADP+Ⓟ_i+H^+$$

In this Unit, we shall refer to ATP and related molecules as 'group transfer substances' in order to imply this role.

SAQ 1 (*Objective 3*)

Classify each of the following statements in relationship to the role of ATP in the cell as most probably

(a) true
(b) false
(c) neither true nor false but an analogy.

1 The free energy of hydrolysis of ATP is of the order of -31 kJ mol^{-1}.

2 ATP is found in practically all living systems.

3 During glucose oxidation, ATP is synthesized from ADP.

4 During glycogen synthesis, ATP is synthesized from ADP.

5 Many cell reactions are 'primed' by a phosphoryl group.

6 The terminal phosphate bond of ATP is unique in character.

7 ATP is involved in muscular contraction.

8 It is possible to regard the cell, for the purpose of thermodynamic analysis, as a closed system.

9 Equilibria are not normally reached in cell reactions; a steady state is the normal condition of the living cell.

10 ATP is the energy currency of the cell.

(Answers on p. 40.)

3.3 Group Transfer Molecules

Study Comment

This section is a review of group transfer molecules. You are not expected to recall individual formulae, but you should remember the broad categories of substances which can serve as such substances.

Among the general classes of compounds coming into the category of group transfer molecules, we may consider in particular two: phosphates and thioesters.

3.3.1 Phosphates

1 *Nucleoside phosphates*

ATP and the three related substances, guanosine triphosphate (GTP), cytosine triphosphate (CTP) and uridine triphosphate (UTP) all come into this category. There are other related coenzymes present in the cell, such as NAD and NADP (S100[15]) which do not function as phosphorylating agents despite their nucleoside component. The formula of ATP is written out in full as shown in Figure 1; during enzyme reactions, ATP can be split at (*a*), (*b*) or (*c*).

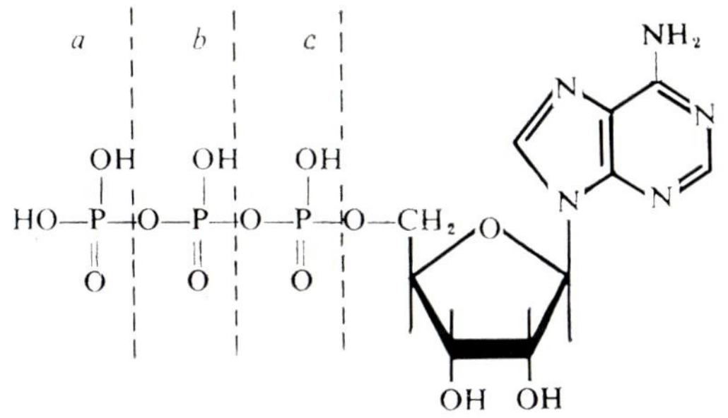

Figure 1 adenosine triphosphate (ATP)

nucleoside phosphates

If the split occurs at (*a*) then ATP is acting as a phosphorylating agent, at (*b*) as a pyrophosphorylating agent, and at (*c*) as an adenylating agent; in the last case, the whole adenine and ribose moiety is transferred to some receptor compound.

Inorganic polyphosphates can under certain circumstances also act as energy-linked coenzymes, and they may react with ADP to give ATP and Ⓟ$_i^-$; in both these cases the relevant bonds are the phosphate-phosphate anhydride bonds.

It may be noted that, contrary to the popular beliefs engendered by the 'high energy' concept, ATP is quite a stable molecule, even in neutral solution at 37°, though it is fairly rapidly hydrolysed if the pH of the solution drops.

2 *Amidine phosphates*

Another class of phosphates, the amidine phosphates, have the general formula

$$R_2N-\overset{\overset{\displaystyle NH}{\|}}{C}-NH-Ⓟ$$

An example with which you are already familiar from S100 Units 15 and 16 is creatine phosphate

$$CH_3N(CH_2COOH)-\overset{\overset{\displaystyle NH}{\|}}{C}-NH-Ⓟ$$

amidine phosphates

which participates in the Lohmann reaction between creatine and ATP. Another example is the substance arginine phosphate, which serves the same function as creatine phosphate in the tissues of many species, notably molluscs.

What is the relevant bond broken in phosphoryl transfer in the amidine phosphates?

The nitrogen-phosphorus bond.

3 *Acylphosphates*
Examples are 1,3-diphosphoglyceric acid (to which we return in Section 3.5.1) and acetyl phosphate

acyl phosphates

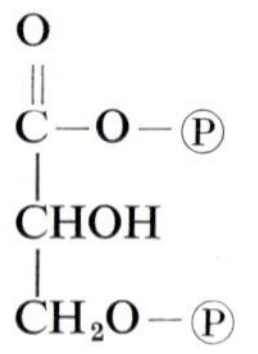

1,3-diPGA

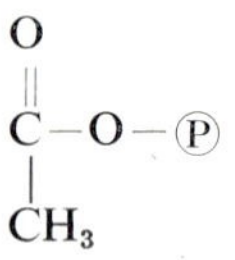

acetyl phosphate

Both will transfer phosphate to ADP.

Which is the bond broken in phosphoryl transfer from acetyl phosphate?

The bond between the oxygen of the carboxyl group and phosphate, thus

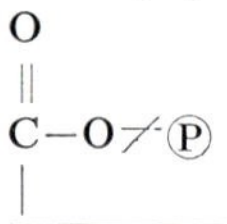

There are two phosphate groups in diphosphoglyceric acid. How many molecules of ADP will one molecule of diphosphoglyceric acid phosphorylate to ATP?

Only one. The second phosphate, attached as

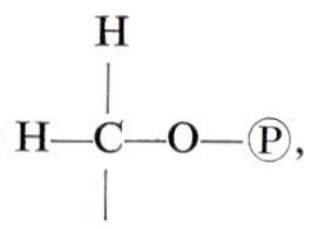

cannot be transferred to ADP (in the inaccurate, old-fashioned terminology, it is a 'low-energy' bond).

In addition to its phosphoryl transfer role, acetylphosphate will act as an acetyl transfer molecule, just as ATP will adenylate. In this role it is perhaps of even more biological significance.

4 *Enolphosphate*
Phosphoenolpyruvate (PEP) is the principal example here:

enolphosphates

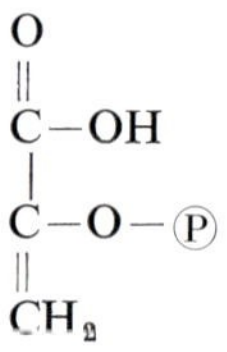

This substance is an intermediate in the glycolysis sequence (see Unit 4) and will phosphorylate ADP.

SAQ 2 (*Objective* 4)

Examine the formulae of these four classes of molecules. What common feature(s) distinguish them from compounds containing phosphate groups of the type of the second, non-transferable group in diphosphoglyceric acid? (Answer on p. 40.)

The $\Delta G^\ominus$ of hydrolysis of the phosphates which can participate in phosphorylation reactions is generally -31 kJ mol^{-1} or greater. By contrast, that of phosphates which cannot is the order of -12 kJ mol^{-1}, although there are intermediate categories, such as phosphates bound to the hydroxyl groups of such amino acids as serine and histidine in proteins (phosphoserine, phosphohistidine).

3.3.2 Thioesters

A quite different class of compounds which participate in energy-linked reactions and can be classed as group transfer molecules depend for their properties not on the presence of phosphate groups but upon sulphur. There is a substance which is midway between the phosphate and the sulphur-containing coenzymes, called adenosine diphosphosulphate, which has a role in some chemotrophic bacteria (S100[16]) but there is an almost universally occurring coenzyme containing sulphur, which is of much greater significance, the substance known as coenzyme A. It was a discovery of Lipmann's in the early 1950s, and derived from his work with acetyl phosphate. He was examining acetyl phosphate not from the point of view of its role in phosphorylation but in acetylation; it was, he felt, an 'activated' form of the acetate molecule. Later, he discovered that the activated acetate was not acetyl phosphate at all but a compound into which it was easily converted, the acetyl-bound form of coenzyme A (Fig. 2).

Coenzyme A

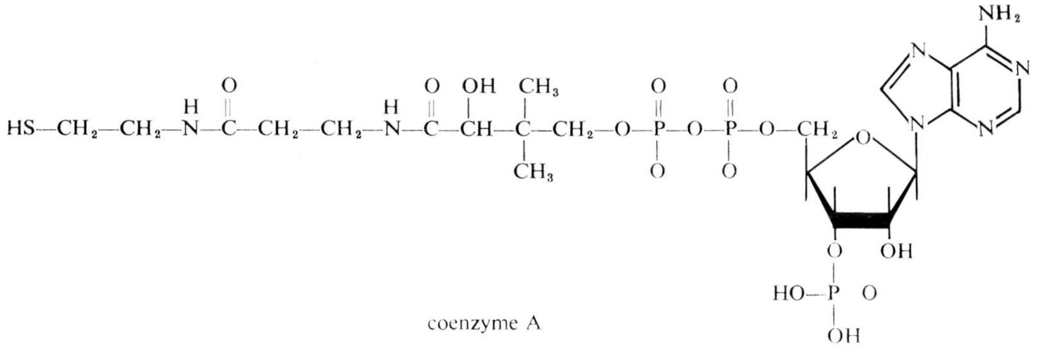

coenzyme A

Part of the coenzyme A molecule is the substance pantothenic acid, one of the B group of vitamins, but the region involved in the acetylation reactions is that terminating in the sulphydryl group—SH. The molecule may be represented as CoASH and in its acetylated form as

$$CH_3\overset{\overset{O}{\|}}{C}-S-CoA \text{ (or } R-\overset{\overset{O}{\|}}{C}-S-CoA)$$

The $\Delta G^{\ominus}$ of hydrolysis of the acetyl-S group is of the order of -25 kJ mol^{-1}, and the molecule finds a role in group transfer reactions similar to that of ATP. In fact, acetylation is a special case of a general class of reactions involving CoA, *acylation* reactions, in which acyl S—CoA, generated in exergonic reactions, is subsequently utilized either in the transfer of acyl groups (as for example in the synthesis of fatty acids, discussed in Unit 4, or of the nervous transmitter molecule acetylcholine, formed from a molecule of acetyl CoA and one of choline—discussed in SDT 286, Unit 1). Alternatively, a molecule of acyl CoA, generated during oxidation, can be utilized for the synthesis of acylphosphate and hence ATP, thus

acylation reactions

$$CH_3\overset{\overset{O}{\|}}{C}-SCoA+Ⓟ_i^-+H^+ \rightleftharpoons CH_3\overset{\overset{O}{\|}}{C}-O-Ⓟ+CoA.SH$$

$$CH_3\overset{\overset{O}{\|}}{C}-Ⓟ+ADP \rightleftharpoons CH_3\overset{\overset{O}{\|}}{C}-OH+ATP$$

This particular reaction occurs in some bacteria. In the following sections of this Unit we will meet other reactions involving the generation of ATP from acyl CoA.

3.4 The Cellular Roles of ATP

Of all the group transfer molecules that were considered in the last Section, the most important is ATP. All living organisms contain ATP (with the exception of viruses, whose 'living' status is a question of definition, of course). The presence of ATP may indeed be taken as almost an index of life—in fact, in considering the development of monitoring devices which could detect a supposed 'biological warfare' attack by an enemy, the use of sensitive ATP-detectors has been proposed, though of course this would not tell us whether the ATP-containing living organisms were pathogenic. ATP-detectors have also been suggested for space missions looking for life on other planets. ATP can be synthesized abiotically (S100[17]), but this presumably does not explain its evolutionary ubiquity. Other group transfer molecules, such as creatine phosphate, are replaced in some organisms by arginine phosphate. A good explanation of 'why ATP?' is still lacking.

Before going on to examine the ATP-generating reactions in more detail, we may pause to consider some of the multitude of cellular reactions which are ATP-dependent.

1 The first class of roles of ATP is in chemical synthesis: (a) in the synthesis of glycogen from glucose (S100[18]), two ATP molecules are converted to two of ADP for each glucose unit added to the glycogen chain; (b) in the synthesis of fatty acids (S100[19]), the equivalent of two molecules of ATP are converted to two of ADP for each 2-carbon fragment added to the fatty acid chain; (c) in the synthesis of protein (S100[20]), probably the equivalent of at least four and maybe more ATP molecules are converted to ADP for each peptide bond formed; (d) in the synthesis of small molecules, such as amino acids, purines, pyrimidines, hormones, nervous transmitters, steroids and the myriads of complex 'natural products' synthesized by plants and bacteria (some of which are shown in the metabolic chart included in S100, Unit 15), more ATP molecules are consumed.

ATP in synthesis

2 The second major class of cellular function involving ATP is that in which the energy transduction is not between one class of chemical substances (e.g. glucose), which is broken down, and another (e.g. protein), which is synthesized, but between chemical energy, as in glucose, and mechanical, heat and informational energy. Thus we may specify an involvement of ATP in:

(a) the maintenance of homeostasis (S100[21]) within the cell. ATP is converted to ADP during the transport of metabolites, such as amino acids, into the cell against a concentration gradient, or in the transfer of the ions of sodium and potassium across the cell membrane in order to maintain the internal high K^+/low Na^+ ionic concentration which is crucial to a variety of cell functions (S22–, Unit 10; SDT 286, Unit 1). One molecule of ATP is converted to ADP for about each three Na^+ ions transported from inside out of the cell. Osmoregulation, the control of water balance in animals, can also be included under this category.

ATP in homeostasis

(b) muscular contractions, in which ATP and creatine phosphate are broken down to ADP and creatine (S100[22] and SDT 286, Unit 4). The probable mechanism for muscular contraction involves the splitting of ATP during the movement of two protein molecules (actin and myosin) relative to one another; the ATP is immediately resynthesized at the expense of creatine phosphate.

muscle and nerve

(c) nervous transmission, both along the axon (S100[21] and SDT 286, Unit 1) and across the junctions, or synapses, between the cells (S100, Unit 18, and SDT 286, Unit 1). ATP is involved in these systems in the restoration of ionic balances after the passage of a nerve impulse and in the synthesis of the chemical transmitters involved at the synapse.

(d) biological light emission. Many small marine plankton organisms, some deep sea fish, and beetles such as glow-worms and fireflies are all capable of emitting light. In the firefly system, an enzyme, luciferase, reacts with a substrate, luciferin, and ATP. The chemical reaction is a light-emitting one, during which ATP is broken down to ADP. One method of estimating the amount of ATP in a tissue sample is to use the luciferin/luciferase system derived from dried crushed firefly tails and count the light flashes emitted when ATP is added (using a scintillation counter—see TV4). The method is sensitive down to the picomol (10^{-12} mol) level of ATP concentration.

light emission

(e) biological 'defence mechanisms'. Organisms such as the electric eel can administer a powerful—even dangerous—electric shock when touched. This shock is delivered by means of an organ known as the electroplax (which in many cases may represent a substantial proportion of the animal's body mass). High voltages are built up by the 'stacking' of cells, each of which maintains ionic differentials across the membrane—in fact, an elaboration of the neuromuscular junctions discussed in SDT, 286, Unit 1. The stacked electroplax cells act as a battery, discharging their potential differences when the shock is administered. The ionic differentials are subsequently restored again by expenditure of cellular ATP.

defence mechanisms

In spite of this multiplicity of cellular function, most cells maintain only relatively small internal concentrations of ATP. In mammals, one of the highest concentrations occurs in brain tissue, where around 1×10^{-6} mol of ATP/g wet weight of tissue is a normal concentration. ADP and AMP levels are normally very much lower. Yet the maximum rate of utilization of ATP in the brain *in vivo* (during electroshock, for instance, when many of the cells are discharging) is of the order of 1 200–2 000 $\times 10^{-6}$ mol of ATP (converted to ADP)/g/hour. These rates cannot be maintained for many minutes, however, because the maximum rate of ATP synthesis from ADP is 250–500 $\times 10^{-6}$ mol ATP synthesized/g/hour. If the maximum rate of utilization was continued for more than a short while, ATP levels would be totally depleted, and ADP and AMP levels would rise. Such rises can indeed be detected under certain circumstances *in vivo*, and quite easily *in vitro*.

ATP levels

Nonetheless, at the maximum resynthesis rate, each molecule of ATP must turn over between 250 and 500 times per hour, while the steady state concentration of ATP is maintained. ATP levels within the cell are thus maintained fairly constant—yet another reason why naïve thermodynamics cannot be applied to the situation.

3.5 The Generation of Group Transfer Molecules

Having surveyed some of the cellular functions of ATP, we may now turn to look in more detail at the reactions which generate it within the cell.

There are a number of main types of reactions generating group transfer molecules, of which four can be listed for detailed consideration:

1 Phosphorylation of glyceraldehyde-3-phosphate.

2 Dehydration of 2-phosphoglyceric acid.

3 Cleavage of C—S, C—C and C—N bonds.

4 Electron transport oxidative phosphorylation.

Of these the fourth is the most important; it is the only one dependent on the presence of oxygen and, under normal conditions, when oxygen is present in the aerobic organism (S100[23]), it accounts for more than 70 per cent of the total ATP generation. In eucaryotes, it occurs, of course, within the mitochondria. Of the remaining three types, (1) and (2) occur outside the mitochondria; (3) may occur either within *or* outside it. We shall first consider some examples of the first three types, before examining in more detail the mechanisms of oxidative phosphorylation itself. In general, those non-mitochondrial syntheses, in which ATP is the group transfer molecule generated, which involve the direct interaction of substrate with ADP, are described as substrate level phosphorylations.

Before considering the reactions in greater detail, however, we will need to note that, in many of the reactions we will have to consider, a number of additional coenzymes occur, some of which you have not previously met. You will need to know these only by their initials, for we are not interested here in the intimate molecular mechanisms of their interconversions; their formulae are listed in Appendix 1.

These coenzymes are as follows:

(a) NAD; nicotinamide adenine dinucleotide, which you have already met. It is capable of reversible oxidoreduction, represented as

NAD

$$NAD + 2[H] \rightleftharpoons NADH_2$$

as we showed it in S100 and in the convention we continue to use here, as it is compatible with the formulation in your set book. More accurately, the reaction should be represented as

$$NAD^+ + XH_2 \rightleftharpoons NADH + H^+ + X$$

where X is the substrate being oxidized.

(b) FP; flavoprotein, which is capable of reversible oxidoreduction, represented as

$$FP + 2[H] \rightleftharpoons FPH_2$$

FP

(c) TPP; thiamine pyrophosphate, a form of the B group vitamin thiamine (see Unit 2) which participates in reactions involving the removal of CO_2 from a molecule (decarboxylation).

TPP

(d) LSH; lipoic acid, a coenzyme containing two sulphur groups which can exchange reducing equivalents reversibly with coenzyme A (CoA.SH). In reaction it is best represented as

LSH

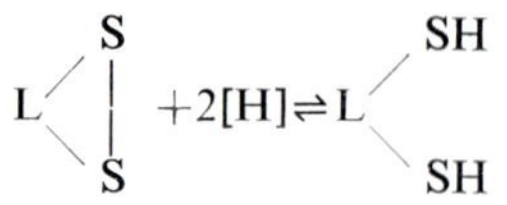

lipoic acid dihydrolipoic acid

3.5.1 Phosphorylation of glyceraldehyde-3-phosphate

Substrate level phosphorylation is said to occur when ATP (or another nucleotide triphosphate like GTP) is generated from ADP directly in the course of an enzyme reaction involving the substrate of a metabolic path; this distinguishes it from phosphorylation of ADP at the expense of coenzymes of the pathway, such as in the oxidation of $NADH_2$ to NAD, etc.

substrate level phosphorylation

You have already come across an example of substrate-level phosphorylation. Name it.

Substrate level phosphorylation occurs during the glycolytic sequence, in the conversion of glyceraldehyde-3-phosphate to 3-phosphoglyceric acid (*Chemistry of Life*, p. 125).

Look back at reactions (6) and (7) on p. 125 in *Chemistry of Life*. Ignoring the 'energy-rich' terminology, note that ATP is generated directly by phosphorylation from a reaction intermediate, 1,3-diphosphoglyceric acid; note also that in reaction (6) $NADH_2$ is generated. You will not be surprised to learn that the two reactions conceal a greater complexity. In fact, the reaction mechanisms have been studied in some considerable detail, since the enzyme glyceraldehyde-3-phosphate dehydrogenase was first crystallized, by Warburg and Christian, in 1939. Racker has shown that both reactions occur on the same enzyme, which contains sulphur groups and also three molecules of bound NAD per molecule of enzyme; we can represent it as

If we represent the non-relevant part of the substrate molecule, $-CH_2O$Ⓟ. CHOH, as R, then the reaction sequence can be written as follows:

(1)

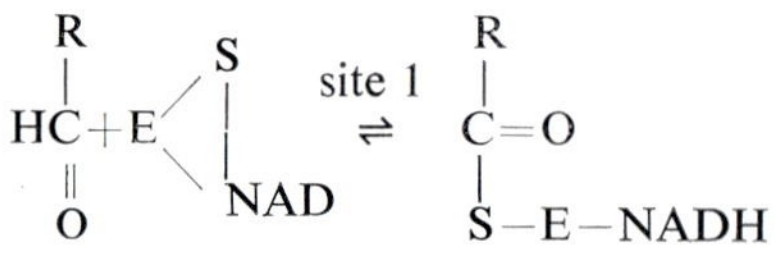

(2)

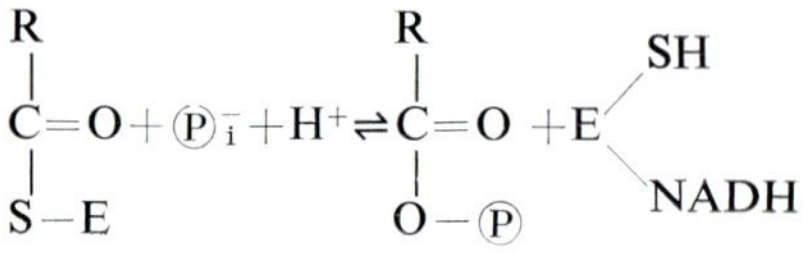

The sulphydryl group of the enzyme and the bound NAD have both now been reduced. Splitting of the 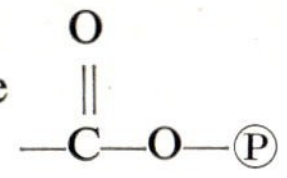 moiety, it will be recalled from Section 3.2 can be coupled with the phosphorylation of ATP, and this reaction occurs on the second site of the enzyme:

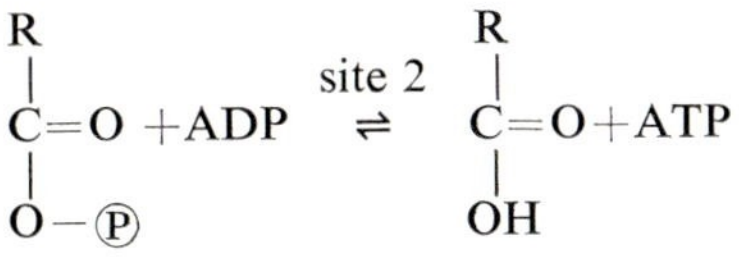

ATP and the product, 3-phosphoglyceric acid, can now be released and the enzyme in its original form can be restored by transferring the hydrogen bound to it to NAD in its oxidized form in the cell cytoplasm.

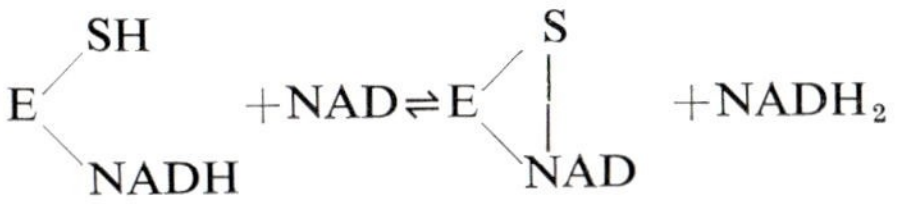

All of the stages in these reactions can be studied and the intermediates, including the bound enzyme complexes, isolated. This makes the reaction system rather a good model for more complex ATP generating systems (see Section 3.8).

3.5.2 Dehydration of 2-phosphoglyceric acid

You have also met this reaction already. In what context, and what are the enzymes responsible? 2PGA

See pp. 126–7 in *Chemistry of Life*. The enzymes are enolase and pyruvic phosphokinase, and the reaction is another of the glycolytic sequence.

3.5.3 Bond cleavage

We now come to the most important of the group transfer molecule-generating reactions other than electron transport; these are those in which bond cleavage occurs, particularly of C—C and C—S bonds. Typical of such reactions is the one which occurs at the link-point between glycolysis and the citric acid cycle; it is the oxidative cleavage of the pyruvic acid molecule CH_3COCO_2H to yield carbon dioxide CO_2, H_2O and an acetyl fragment, CH_3.CO. (*The Chemistry of Life*, pp. 128–30.) Acetyl CoA and reduced $NADH_2$ are generated during this reaction sequence, which is analogous to and a model for the oxidative decarboxylation of α-oxoglutaric acid (*The Chemistry of Life*, p. 133), during the citric acid cycle, and also of the cleavage of fatty acids during their oxidation (*The Chemistry of Life*, pp. 139–43).

In this type of oxidative decarboxylation, a complex reaction sequence involves a sulphur-containing enzyme complex of three separate enzymes, E_1, E_2 and E_3 (see the picture in S100[24]) and the coenzymes, TPP, LSH and CoASH. (See also Unit 1 of this Course.) The full reaction sequence (do *not* try to recall it) can be represented as follows:

pyruvate decarboxylase

(1) $E_1-TPP+CH_3COCO_2H \rightleftharpoons E_1-TPP-CHOCH_3+CO_2$

(this is a straightforward C—C bond cleavage; carbon dioxide is released and the resultant acetaldehyde fragment, bound to TPP (by way of a C—C bond) can be 'activated' and participate in group transfer reactions. E_1 is the enzyme pyruvic carboxylase).

(2) 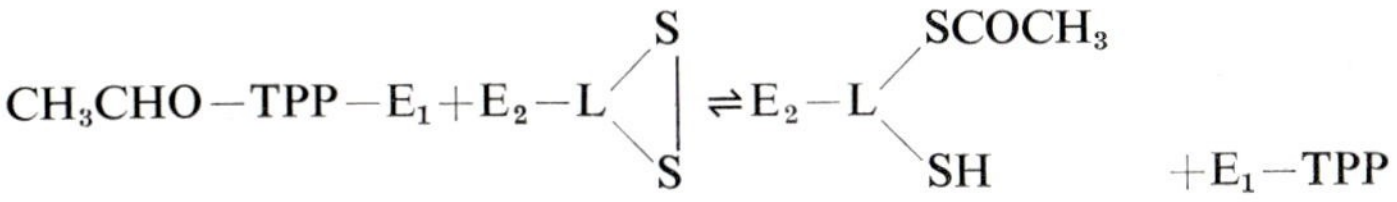

(Here the acetyl fragment is transferred to the lipoic acid containing enzyme E_2, known either as aldehyde dehydrogenase or lipoic acid reductase. $TPP-E_1$

is released in its original form. One of the two sulphurs of the lipoic acid is reduced at the expense of the acetaldehyde fragment which is converted to an acetyl fragment. The C—C bond linking the acetyl fragment to TPP is converted to a C—S bond linking it to the lipoic acid.)

(3) $$E_2{-}L\begin{matrix}\diagup SCOCH_3\\ \diagdown SH\end{matrix} + CoA.SH \rightleftharpoons E_2{-}L\begin{matrix}\diagup SH\\ \diagdown SH\end{matrix} + CoASCOCH_3$$

(The lipoic acid on E_2 has now been fully reduced at the expense of coenzyme A and the acetyl group is now attached to the activated, oxidized CoA molecule. We will return to the fate of this acetyl CoA shortly. The reduced E_2-lipoic acid complex must now be oxidized; the oxidation occurs at the expense of the next coenzyme, FAD.) **acetyl CoA**

(4) $$E_2{-}L\begin{matrix}\diagup SH\\ \diagdown SH\end{matrix} + E_3{-}FAD \rightleftharpoons E_2{-}L\begin{matrix}\diagup S\\ \quad |\\ \diagdown S\end{matrix} + E_3{-}FADH_2$$

(5) $$E_3{-}FADH_2 + NAD \rightleftharpoons E_3{-}FAD + NADH_2$$

All the enzymes and coenzymes have thus been returned to their original oxidized form, and one molecule of $NADH_2$ generated from one of NAD. The enzyme E_3 is dihydrolipoic acid dehydrogenase.

We can now consider the fate of the acetyl CoA.

State one reaction in which acetyl CoA participates?

In the citric acid cycle, acetyl CoA combines with the 4-carbon oxaloacetic acid to yield the 6-carbon citric acid and reduced CoA again.

(6) $$\begin{array}{c} CH_3COSCoA\\ +\\ O{=}\underset{\displaystyle CH_2CO_2H}{\underset{|}{C}}{-}CO_2H \end{array} \rightleftharpoons HO{-}\overset{\displaystyle CH_2CO_2H}{\overset{|}{\underset{\displaystyle CH_2CO_2H}{\underset{|}{C}}}}{-}CO_2H + CoASH$$

The acetyl CoA molecule is sufficiently activated, because of the CoA complex, that it can participate in the C—C bond formation involved in this reaction; subsequently, of course the citric acid is oxidized through the citric acid cycle. (See *Chemistry of Life*, p. 132. The enzyme responsible is the citric condensing enzyme; see also Unit 4 of this Course. Another set of reactions including CoA, not discussed here, is that of fatty acid synthesis.)

However, the utilization of acetyl CoA for C—C bond formation, by-passing as it does the generation of ATP, is only one possible fate. In bacteria, for example, acetyl CoA is involved in ADP phosphorylation to ATP, thus

(7) $$CoASCOCH_3 + ⓟ_i^- + H^+ \rightleftharpoons CoASH + CH_3COⓟ$$

This substance is acetyl phosphate, and will be recalled from Section 3.2 above. It can participate in the reaction

(8) $$CH_3COⓟ + ADP \rightleftharpoons CH_3CO_2H + ATP$$

The enzyme for reaction 7 is transacetylase; for reaction 8, acetokinase. The transacetylase enzyme is tightly bound, in bacteria, to enzyme E_3, the lipoic acid reductase of equation 2. In bacteria, the sum of this complex reaction sequence is thus

(9) $$CH_3COCO_2H + ADP + NAD + ⓟ_i^- + H^+ \\ \rightleftharpoons CH_3CO_2H + CO_2 + ATP + NADH_2$$

The pyruvic acid has been decarboxylated and oxidized to acetic acid, one molecule of $NADH_2$ and one of ATP being produced.

It is scarcely surprising that an enzyme sequence of this complexity demands a multienzyme complex, the pyruvic decarboxylase/dehydrogenase complex, to carry it out with maximal efficiency. The complex shown in S100[24] contains 1 unit of the reductase transacetylase (with a molecular weight around 1.6×10^6) compared with 12–14 units of the carboxylase (enzyme E_1), molecular weight 1.8×10^5, and 6–8 units of the dihydrolipoic acid dehydrogenase, molecular weight 1.1×10^5.

This multienzyme complex is thus a good example of the type of higher order organization of proteins discussed in Unit 1. But why is so complex an enzyme sequence required for these reactions? The answer is in itself a good introduction to the problems of mitochondrial oxidative phosphorylation. There are three principal points. The first is that for maximal efficiency in a reaction sequence free diffusion of substrates and enzymes is undesirable; their close coupling will obviously ensure that the reactions occur in an orderly sequence with maximal efficiency. The second is that, if the maximal number of group transfer molecules are to be synthesized during the single oxidative decarboxylation reaction, the reaction needs to be broken down into steps in each of which just one of the group transfer molecules can be converted to its activated form; the third point is that in their activated form they may be sensitive to hydrolysis by water. The least efficient reaction, from the point of view of the cellular economy, is that of:

$$ATP + H_2O \rightleftharpoons ADP + Ⓟ_i + H^+$$

This reaction can be avoided if hydrogen and hydroxyl ions are excluded from the sites at which ATP synthesis is occurring. This exclusion can occur most easily if the sites are in the interior of large protein or lipoprotein complexes (Units 1 and 2). A multienzyme complex can provide such an environment. So, too, as we shall see, can the mitochondrion. In a way, then, ATP and $NADH_2$ generation on the pyruvic dehydrogenase complex is a model for mitochondrial events.

SAQ 3 (*Objectives* 1, 6)

The phosphorylation of ATP during the enolase-pyruvic phosphokinase reaction is an example of:

(a) oxidative phosphorylation,

(b) substrate level phosphorylation,

(c) C—C bond cleavage,

(d) none of these,

for the following coenzymes:

(a) TPP; (b) FAD; (c) NAD; (d) ATP; (e) lipoic acid; (f) CoA; (g) ADP.

SAQ 4 (*Objectives* 1, 4)

Which contain sulphydryl groups?

SAQ 5 (*Objective* 6)

Which participate in the glyceraldehyde-3-phosphate dehydrogenase reaction?

SAQ 6 (*Objective* 6)

Which participate in the pyruvic dehydrogenase reaction?

SAQ 7 (*Objective* 6)

Which are directly involved in decarboxylation reactions?

SAQ 8 (*Objective* 6)

Which of the following descriptions is the most complete statement of the function of the pyruvic dehydrogenase complex:

(a) The phosphorylation of 2-phosphoglyceric acid.

(b) The synthesis of ATP from ADP and phospholenol pyruvate.

(c) The decarboxylation of pyruvic acid.

(d) The oxidative decarboxylation of pyruvic acid.

(e) The synthesis of citric acid from pyruvic acid and oxaloacetic acid.

(f) The reduction of NAD to $NADH_2$.

(Answers on p. 40).

3.6 Oxidative Phosphorylation and the Mitochondrion

3.6.1 The structure of the mitochondrion

We may now turn to the organelle whose properties will occupy us for practically the whole of the rest of this Unit, the mitochondrion, 'power-house of the cell'. We are concerned with the mitochondrion from the point of view of its role in energy transduction, its main cellular function. We need to consider both the function of the mitochondrion, that is, how it synthesizes ATP from ADP, and its structure in relation to that function. In making this study, we must be very conscious that it is much harder to examine the properties of the mitochondrion in relation to oxidative phosphorylation than it is to study the action of glyceraldehyde-3-phosphate dehydrogenase or even pyruvic dehydrogenase. We shall find a very close relation between results obtained, interpretations put upon them, and the methods used in investigation—much closer than in almost any other problem of biochemistry. Techniques will therefore occupy us not a little. But, first, you should revise and expand your knowledge of mitochondrial structure. To do this, turn to *Cell Structure and Function*, Chapter 13, and read from the beginning of the Chapter, on p. 310, to p. 314. Much of the description should be familiar to you—you saw mitochondria being isolated in S100[25]. Of the two enzymes referred to on p. 312, you should recall cytochrome oxidase from S100[26]. 'Succinoxidase', an abbreviated form of 'succinic oxidase' is the name given to the entire complex of dehydrogenase and electron transport chain which oxidizes succinic acid, a citric acid cycle intermediate, to fumaric acid. Note that *Cell Structure and Function* refers to the citric acid cycle as the Krebs cycle. You will know why (from S100[27]) this is one of the alternative names for this metabolic sequence. Table 13.1 (p. 315) is a good example of the use of specific activity data to demonstrate an enzyme purification—though this time subcellular—as in the hexokinase experiment of TV 1, 2 and 3.

mitochondrial structure

Now answer *SAQ*s 9–11 (Answers on p. 40).

SAQ 9 (*Objective* 9)

Arrange the following subcellular organelles in decreasing order of particle weight:
mitochondrion; ribosome; lysosome; nucleus.

SAQ 10 (*Objective* 9)

How many types of membrane are there in the mitochondrion?

SAQ 11 (*Objective* 9)

Which of these descriptions is correct:

(a) Within the mitochondrion, there is a lipoprotein matrix made by the infolding of the inner membrane, within which there is a liquid-filled space in the cristae.

(b) Within the mitochondrion, there are lipoprotein cristae made by the infolding of the inner membrane, within which there is a liquid-filled space known as the matrix.

Check your answers before continuing (p. 40).

3.6.2 Oxidative phosphorylation; the development of the problem

What is the evidence that links oxidative phosphorylation, as well as the citric acid cycle, to the mitochondria? The section of *Cell Structure and Function* headed '*Mitochondria as energy converters*' (pp. 315–16) provides some of it. In reading it, note that the authors refer to α-oxoglutarate as α-keto glutarate. The ratio of the number of moles of phosphorus, converted to ATP, to oxygen consumed is known as the *P/O ratio*. It is the process of ATP production during the oxidation of substrate by way of the carriers, which is known as oxidative phosphorylation. By the 1950s, work by Lehninger, Lardy, Green and others,

P/O ratio

had shown that the P/O ratio of the individual oxidation steps of the cycle were as follows:

Table 1

Step	*P/O ratio*
pyruvic→acetyl CoA	3
isocitric→α-oxoglutaric	3
malic→oxaloacetic	3
α-oxoglutaric→succinic	4
succinic→fumaric	2

The P/O ratio of four in the α-oxoglutaric acid→succinic acid step includes one molecule of ATP generated by the oxidative decarboxylation of α-oxoglutaric acid in a manner analogous to that of pyruvic acid described in Section 3.5.3.

These ratios can be obtained, with care, in mitochondrial preparations free of almost any other cellular contamination, while no other isolated cell organelle is capable of carrying out oxidative phosphorylation. This is good evidence for the function of the mitochondrion in eucaryotic organisms. Bacteria, on the other hand, do not possess mitochondria in the form discussed here, although they do possess membrane-bound enzymes which can conduct oxidative phosphorylation. More evidence as to the role of the mitochondria comes from the fact that tissues with a major ATP requirement, such as muscle, kidney and secretory cells, are richer in mitochondria than those without such requirements. Another piece of evidence is that when yeast is grown anaerobically, oxidative phosphorylation cannot, of course, occur, and normal mitochondria cannot be seen in the cells. When the anaerobic yeast is switched to grow in the presence of oxygen, mitochondria soon appear within the cells which become capable of oxidative phosphorylation. Knowing that oxidative phosphorylation occurs within the mitochondrion is a good start, but how can one begin to unpick the mitochondrion and its enzymes far enough to understand the detailed mechanisms involved? The problem resolves itself essentially into four portions: (a) the localization of the citric acid cycle enzymes; (b) the nature of the electron transport chain; (c) the mechanism of oxidative phosphorylation itself; (d) the nature of the systems which control and regulate mitochondrial events. Problem (a) is almost universally accepted as being solved; all citric acid cycle enzymes (with the exception of succinic dehydrogenase, SDH) are soluble and can be released by breaking the mitochondrion by sonication (exposure to high frequency sound) or by freezing and thawing or osmotic shock (exposing the mitochondria to hypo-osmotic conditions, such as distilled water). The citric acid cycle enzymes are thus regarded as being located, in solution, in the mitochondrial matrix. Substrates for the enzymes must diffuse through the mitochondrial membranes in order to be metabolized, and waste products, mainly CO_2, diffuse out again. Problem (c) we will return to below (Section 3.8), and problem (d) must wait until Unit 4 for a more detailed treatment.

ATP and the mitochondrion

3.7 The Electron Transport Chain

The problem of the electron transport chain is but one example of a general biochemical problem. A series of tightly coupled reactions occur, catalysed by an insoluble lipoprotein preparation. The reactions occur in sequential order and are dependent on the structural integrity of the lipoprotein complex, for if it is solubilized with detergents, they can occur only at very much reduced rates.

The problem is to determine:

(a) the nature of the individual components of the chain;

(b) the sequence in which they operate;

(c) the reaction mechanisms involved.

These problems are rather intractable and more than twenty years of intensive research in many laboratories has failed to provide anything but partial answers.

A number of general methods of approach have been adopted, which we shall consider briefly in turn. These are:

(a) the isolation of individual chain components;
(b) the use of submitochondrial complexes;
(c) the use of inhibitors of the reactions;
(d) the use of stopped-flow and steady-state methods;
(e) the use of artificial electron acceptors.

3.7.1 The isolation of individual chain components

This approach regards the problem of the chain essentially as one in enzymology: particular proteins or other components of the lipoprotein complex are purified by variants of the classical techniques you read about in Unit 1; for example the enzymes succinic dehydrogenase or $NADH_2$ dehydrogenase or cytochrome oxidase. Individual coenzymes, such as NAD or coenzyme Q can be isolated similarly. Only one of the chain components, cytochrome c, is readily soluble in water and is easily extracted into solution, where it can be purified and studied in isolation.

SAQ 12 (*Objectives* 2, 7)

Which of the following criteria are of relevance in deciding whether or not a substance isolated from the mitochondrion is a component of the electron transport chain? Which are both *necessary* and *sufficient* criteria; that is, contain a complete specification of the characteristics of the intermediate?

the Slater criteria

1 It is present in large quantities compared to other substances already known to be part of the chain.

2 It is present in stoichiometric quantities compared to other substances known to be part of the chain.

3 It is oxidized and reduced at rates appropriate to the known rate of oxidation and reduction of substances known to be part of the chain.

4 It oxidizes reduced $NADH_2$.

5 It reduces cytochrome a.

6 Removal of the alleged component inactivates all or part of the chain.

7 Addition of the purified alleged component inactivates all or part of the chain and increases its oxidoreduction rate.

8 Removal of the alleged component inactivates all or part of the chain; its addition reactivates it.

Check your answers before continuing (p. 40).

The necessary and sufficient criteria, as defined in the answers to this *SAQ*, were specified in the 1950s by Slater, and are of general interest in the analysis of other situations of biochemical importance. You should be able to recall them and recognize their general validity. In fact, on the basis of such criteria, and of isolation and purification studies, about 25 per cent of the protein of the mitochondrion appears to be involved in electron transport and oxidative phosphorylation as carrier or coenzyme. The remainder of the protein is probably either associated with the maintenance of membrane structure, or composes the enzymes for the many other reactions of which the mitochondrion is capable.

If it seems at first sight surprising to you that components can be removed and replaced again from a highly ordered structure such as the electron transport chain, you are right, it is surprising! Almost certainly the conformational reasons why this is possible relate to the sort of features discussed in Unit 1 with respect to the quaternary structure of proteins. These are coming to be seen as reflecting a biochemical principle of great generality and considerable importance so far as understanding the cell is concerned.

When the Slater criteria are applied, the individual chain components which are present in significant quantities in the mitochondrion can be taken to include the following:

1 The NADH dehydrogenase flavoprotein (FP).
2 The succinic dehydrogenase flavoprotein (SDH).
3 The quinone, coenzyme Q (CoQ).
4 Five cytochromes, a, a_3, b, c and c_1.
5 Cu^{2+} ion, associated with cytochromes a and a_3.

6 A certain amount of iron *not* bound to the haem groups of the cytochromes, some of which is associated with NADH dehydrogenase and SDH; it is known as 'non-haem iron'.

chain components

In what follows, succinic dehydrogenase will be referred to as SDH, NADH dehydrogenase (which contains a flavoprotein prosthetic group) as FP. The non-haem iron seems to be an important component of as yet uncertain role. Note the association of Cu^{2+} with cyt a_3; the copper is involved in its cytochrome oxidase activity (see below). Actually the proportion of individual components of the chain varies slightly from species to species and tissue to tissue; thus rat heart mitochondria have, on a molecular basis, five times as much, and beef heart mitochondria seven times as much coenzyme Q as cytochrome a.

In general, if a component varies in relative concentration markedly from species to species or from tissue to tissue it is probably an indication either that it is not directly involved in the chain or that, in addition to its involvement, it has some other role. NADP, which is present in mitochondria, has a very varied relative concentration, for instance (thus violating the 'stoichiometric criterion') and is not on the direct path of electron transport. On the other hand, for CoQ, which also varies in concentration, it is probably the case that a proportion of the CoQ present in the mitochondrion is in the direct path of electron transport, while the remainder may lie on some side path.

3.7.2 Submitochondrial complexes

This approach is an attempt, not to purify individual chain components, but instead to gently fractionate the mitochondrion into smaller particles, each of which contains some, but not all, of the components and will therefore carry out a particular fraction of the full sequence of mitochondrial oxidative phosphorylation. Historically, the approach goes back to the very beginning of work with mitochondria, by Keilin and Hartree in the 1920s and 1930s. More recently the techniques have been exploited by Green and by Racker; they involve subjecting the mitochondrion to a variety of dissociation treatments, such as sonication, mechanical disruption, the use of agents which bind inorganic ions such as Ca^{2+} and Mg^{2+} (so-called chelating agents), ionic and non-ionic detergents, etc. These procedures work, presumably, by disrupting the structural lipoprotein in which the carriers are embedded. Detergents, in particular, solubilize and extract lipid. Some examples of the appearance of the fragments which can be obtained are shown in the electron micrograph of Figure 13.5 in *Cell Structure and Function* and you should read the account of the particles given there, from the bottom of p. 323 to p. 326.

ETP and PETP

Note that complexes can be obtained which contain the complete capacity for oxidative phosphorylation, like miniature mitochondria (phosphorylating electron transport particles, PETP), or which can transport electrons but have lost their phosphorylative capacity (electron transport particles, ETP). In addition, subfractionation of the ETP will produce at least four separate complexes. Complex I will oxidize reduced $NADH_2$ at the expense of CoQ; Complex II will oxidize succinic acid at the expense of CoQ; (both Complexes I and II contain non-haem iron); Complex III will oxidize reduced CoQ at the expense of cyt c, but will not oxidize $NADH_2$ or succinic acid. It contains cytochromes b and c_1. None of these three complexes can utilize molecular oxygen. However, another such particle, Complex IV, which contains cyt a, cyt a_3 and Cu^{2+} will oxidize cyt c at the expense of molecular oxygen. In addition, and most intriguingly, it is possible to recombine two or more of the complexes by the addition of lipid extracts and cytochrome c (which, because of its solubility, is readily extracted during the isolation of the complexes). When all four complexes are combined, a fully reconstituted ETP is obtained. Presumably, this recombination depends on higher order conformational relationships such as those discussed in Unit 1.

properties of submitochondrial complexes

SAQ 13 (*Objectives* 8, 9)

The description of the properties of the four complexes given in the preceding paragraph should enable you to reconstruct a complete sequence for the relationship of the four complexes to the ETP. Draw each complex in a box, label the components within it, and assemble the four boxes as a flow diagram for the ETP. Then compare your diagram with ours (p. 40).

3.7.3 Inhibitors of electron transport

The rationale here is to find an inhibitor which blocks, for example, the reduction of oxidized CoQ, but will still permit the oxidation of reduced CoQ added to the system. The reactions which will still occur in the presence of the inhibitor must occur *between* CoQ and oxygen; their position in the electron transport chain can thus be determined. On the other hand, the components of the chain towards the substrate end will remain reduced in the presence of the inhibitor; whether individual components are in their oxidized or reduced form can be determined, as you know, by using a spectrophotometer.

There are five sites along the chain at which the inhibitors may act, in experiments with isolated mitochondria or PETP. You need not try to recall the names of the individual inhibitors, but you should now consider the evidence concerning the sites.

1 In the presence of barbiturates* (the components of many sleeping tablets), such as amytal, the oxidation of $NADH_2$ is blocked. All other oxidations, including that of succinic acid, proceed unimpaired.

2 In the presence of rotenone, the oxidation of $NADH_2$ and of some of the non-haem iron is blocked. All other oxidations proceed unimpaired. **sites of inhibitor action**

3 In the presence of thenoyltrifluoroacetic acid, the oxidation of succinic acid and of some of the non-haem iron is blocked. All other oxidations, including that of $NADH_2$, proceed unimpaired.

4 In the presence of the antibiotic antimycin,* the oxidation of succinic acid or $NADH_2$ is blocked. CoQ and cyt b are reduced, but added cyt c can be oxidized.

5 In the presence of ions such as that of cyanide (CN^-) or sulphide (S^{2-}), or substances like carbon monoxide (CO), all the carriers are in their reduced forms; neither $NADH_2$, succinic acid or reduced cyt c can be oxidized.

SAQ 14 (*Objectives* 8, 9)

On the basis of this data and the evidence based on the use of submitochondrial particles, draw a flow chart for the electron transport chain indicating the sites of action of the inhibitors. (Answer on p. 40)

3.7.4 Artificial electron donors and acceptors

The principle of using artificial electron donors and acceptors is rather like that of the inhibitors, but in reverse; the chain, instead of being blocked, is allowed to function using an exogenous non-natural substrate, or electron acceptor. These can be used also to 'bypass' segments of the chain blocked with inhibitors, or to act as acceptors for submitochondrial chain fragments. You have, in fact, already used one such artificial electron acceptor if you did one of the Home Experiments associated with S100, Unit 15 (S100[28]). In that experiment a solution of the dye methylene blue was added to a yeast suspension. Methylene blue is readily reduced and its reduced form is colourless; it will accept electrons from chain components, and can therefore substitute for atmospheric oxygen in mitochondrial oxidations. Other substances which will accept electrons include ferricyanide; electron *donors* include ascorbic acid (vitamin C, present in citrus fruits). A lipid-soluble substance known as menadione (vitamin K_3), related to CoQ, will both donate and accept electrons: it can thus act as a bypass to the chain. The sites of action of some of these components are shown in Figure 3. **electron donors and acceptors**

* *Note that the fact that these substances act like this in* in vitro *systems does* not *necessarily mean that this is their physiological mode of action. In the artificial* in vitro *system, the barbiturates, antibiotics or other inhibitors are often present in concentrations several orders of magnitude higher than could be the case after physiological administration to a human or experimental animal.*

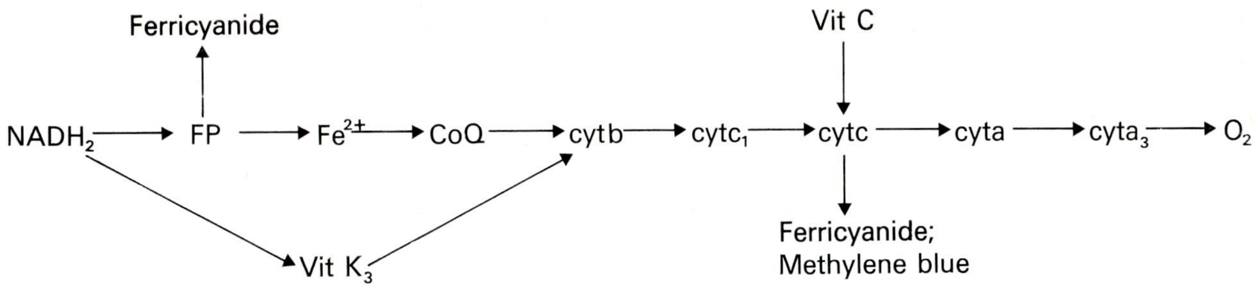

Figure 3

3.7.5 Stopped-flow and steady-state methods

These techniques, pioneered by Britton Chance in the 1950s, are mentioned in the first paragraph on p. 320 in *Cell Structure and Function.* They allow the simultaneous measurement of the state of oxidation or reduction of all the carriers in the chain, and are especially valuable for work with oxidative phosphorylation. The technology is complex and the details need not concern you, but essentially you can envisage a mitochondrial preparation being assayed in a sealed system similar to the tonometer for haemoglobin shown in TV 3. If there is substrate but no oxygen present, all the carriers will be in their reduced state. When a small amount of oxygen is added, there is a sequential conversion of the carriers to their oxidized form, starting with cyt $(a+a_3)$ and ending with the flavoprotein dehydrogenases and NAD; as oxygen is used up, so the carriers become reduced once more. The oxidations and reductions are sequential, and the sequence reveals the order of the carriers along the chain. In fact, even in the presence of unlimited oxygen and the fully active preparation, not all the carriers are in their fully oxidized state; there is, for instance, a gradation between cyt a_3, which is about 87 per cent oxidized, and cyt c which is 65 per cent oxidized. The technique can be used in the presence of inhibitors and artificial acceptors as well. When changes occur in the oxidation/reduction state of the carriers, they are not all in the same direction: thus with limited oxygen, the degree of reduction of cyt a_3 will *increase* while that of the other components will *decrease*. This is called a *crossover site*, and the analysis of these crossover points led Chance to formulate a theory that the crossover site is the site at which phosphorylation occurs during electron transport—an aspect to which we return below.

crossover theorem

3.7.6 The electron transport chain

In the discussion in Sections 3.7.1 to 3.7.5 we have several times described versions of the electron transport chain, each time emphasizing different features. You should by now be familiar with the principal and generally agreed carriers, and you will find a summary account of the chain in *Chemistry of Life*, pp. 112–18, and in *Cell Structure and Function*, pp. 316–18. You should look at both these accounts. Note that there are many ways of drawing the chain, and that concerning the role and location of some minor components, notably the non-haem iron, there is a certain amount of doubt and disagreement. You will probably also find the chain described differently in other accounts you may read; as we suggested at the beginning of this discussion, there is still much controversy in this area; outsiders might regard 'mitochondriologists' almost as 'mitochondriacs' from the tenor of some of their polemics and the obscurity of some of their papers! You will have to become used to uncertainty in this area—especially if you go on to do biochemistry at a higher level; for the purposes of this Course, you should be able to reproduce the principal features of the chain as shown in the set books, noting the main route from $NADH_2$ by way of the FP-linked $NADH_2$ dehydrogenase, and the sidepath from succinic acid via the FP-linked SDH, through CoQ (called CoQ_{10} in *Cell Structure and Function* to distinguish it from other quinones) and the cytochromes to molecular oxygen, and you should appreciate the role of the Fe^{2+} ions of the haem groups of the cytochromes.

electron transport chain

Note the identity of cytochrome a_3 with the enzyme cytochrome oxidase, and note too—not specified in either *Chemistry of Life* or *Cell Structure and Function* —that it is now known that at the terminal involvement of cyt a_3 with O_2, electrons are transferred from the Fe^{2+} of the haem (spelt heme in *Cell Structure and Function*) of the cytochrome to the copper ion Cu^{2+}, also bound to the cytochrome molecule. It is possible that the transfer to molecular oxygen occurs from the copper ion.

All the citric acid cycle dehydrogenases, with the exception of SDH, and practically all the other cellular dehydrogenases—involved, for instance, in fatty acid and amino-acid oxidation—are NAD-linked.

Cytoplasmic dehydrogenations, such as that of glyceraldehyde-3-phosphate discussed above, generate $NADH_2$ free in the cytoplasm. Many of these dehydrogenases have NAD actually bound to the enzyme, which then exchanges 2(H) with free NAD in the cytoplasm. (Many of the dehydrogenases also have the zinc ion Zn^{2+}, as activator.) The free $NADH_2$ must then diffuse to the mitochondrion and cross the mitochondrial membranes before being able to react with the chain. In fact, there is NAD firmly bound to the chain, and in all probability, the free and bound NAD exchange hydrogens as in the reaction

dehydrogenases

$$NADH_2 \text{ (free)} + NAD \text{ (bound)} \rightleftharpoons NAD \text{ (free)} + NADH_2 \text{ (bound)}$$

Also, similar exchanges occur across the mitochondrial membrane between matrix and cytoplasmic NAD, so that cytoplasmic $NADH_2$ never actually enters the mitochondrion and mitochondrial NAD does not leave it. (See also Unit 4.)

NAD itself is thus present in at least four forms:

(a) free in the cytoplasm;
(b) bound to cytoplasmic dehydrogenases;
(c) free in the mitochondrial matrix;
(d) bound to the lipoprotein of the electron transport chain.

The 2(H) of the $NADH_2$ in each of the first three states is eventually transferred to that in the fourth state, whence it passes down the chain; the molecule cannot, of course, be reoxidized by molecular oxygen, though ferricyanide and methylene blue will act as electron acceptors for NAD. In the functioning mitochondrion, the NAD is generally in a fairly oxidized form; ranging from 100 per cent oxidized in brain, through about 90 per cent in heart, kidney and insect flight muscle, to 60 per cent in the liver.

NAD/$NADH_2$

By contrast, the related coenzyme to NAD, NADP, which is not directly involved in the electron transport chain, is generally present in the cell in the reduced form—60 per cent of it is present as $NADPH_2$ in the brain for instance, 91 per cent in the liver. There is much more NAD than NADP in the cell in most tissues, and a relatively small number of the dehydrogenases are NADP-linked; in general, as you will see in Unit 4, NADP is more frequently involved in synthetic rather than catabolic reactions. Among the few NADP-linked dehydrogenases is that for the citric acid cycle intermediate isocitric acid. The role of isocitric dehydrogenase will be considered further in Unit 4.

NADP

3.7.7 Summary of electron transport

The electron transport chain is a set of carrier proteins, embedded in the lipoprotein of the inner mitochondrial membrane and cristae; the proteins carry a set of prosthetic groups capable of being alternately oxidized and reduced; from the substrate end these are flavoprotein, non-haem iron, CoQ, cyt b, cyt c_1, cyt c, cyt a and cyt a_3. One type of flavoprotein-linked enzyme accepts hydrogen from $NADH_2$, to which most of the dehydrogenase systems of the cell are coupled. Another branch line into the chain has a different flavoprotein-linked enzyme which accepts hydrogen from the citric acid cycle intermediate succinic acid (SDH). There are a variety of techniques for the study of the chain; these include isolation of components, submitochondrial particles, use of inhibitors, artificial electron donors and acceptors, and stopped-flow and steady-state methods; the last enables a series of 'crossover sites' along the chain to be identified. The methods are in general concordance concerning the carrier sequence, but points of disagreement or obscurity still exist. Interpretations have varied with methodology and the research school concerned; this is a common feature of active research areas, and demonstrates the extent to which methodology determines results and conclusions. The role of NAD at the entry into the chain is important in terms of the relationship between the mitochondria and the external cytoplasmic enzyme systems.

SAQ 15 (*Objective* 8)

Which of the following dehydrogenases are not linked to the electron transport chain by way of NAD:

(a) Pyruvic dehydrogenase.

(b) Succinic dehydrogenase.

(c) Malic dehydrogenase.

(d) Glyceraldehyde-3-phosphate dehydrogenase.

SAQ 16 (*Objective* 8)

The electron transport chain contains a number of cytochromes. Which of the following best represents the sequence of the cytochromes, from the reducing to the oxidizing end, in the chain:

(a) $a \rightarrow a_3 \rightarrow b \rightarrow c_1 \rightarrow c$.

(b) $c \rightarrow c_1 \rightarrow c \rightarrow a \rightarrow a_3$.

(c) $b \rightarrow c_1 \rightarrow c \rightarrow a \rightarrow a_3$.

(d) $b \rightarrow c \rightarrow c_1 \rightarrow a_3 \rightarrow a$.

(e) $b \rightarrow c_1 \rightarrow c \rightarrow a_3 \rightarrow a$.

SAQ 17 (*Objectives* 1, 8)

Proteins containing flavin as prosthetic group occupy a key position in the electron transport chain. Is this because:

(a) when oxidized they can accept hydrogen from NAD;

(b) when oxidized they can accept hydrogen from succinic acid;

(c) when reduced they can donate hydrogen to NAD;

(d) when reduced they can donate hydrogen to succinic acid;

(e) when oxidized they can be reduced by oxygen;

(f) when oxidized they can accept hydrogen from both NAD and succinic acid;

(g) when reduced they can donate hydrogen to both NAD and succinic acid.

SAQ 18 (*Objective* 1)

Match the items in list A with the appropriate description(s) in list B.

List A	*List B*
cyt a_3	(a) Is a dehydrogenase.
FAD	(b) Is a prosthetic group.
NAD	(c) Is a coenzyme.
Co Q	(d) Is a cytochrome.
Vit K	(e) Is a quinone.
SDH	(f) Is not a component of the electron transport chain.
NADP	(g) Transfers 2 (H).
	(h) Transfers electrons.
	(i) Can be reduced by molecular oxygen.
	(j) Can be oxidized by molecular oxygen.

3.8 Oxidative Phosphorylation

The electron transport chain is a mechanism whereby the energy-producing reactions in which oxygen is reduced by a variety of carriers of reducing energy can be coupled to the synthesis of ATP. As was specified in Section 3, the P/O ratio during $NADH_2$ oxidation is 3, that during succinic acid oxidation is 2. In relationship to oxidative phosphorylation, there are two questions of most interest; the first is: where does this coupling occur in the carrier sequence; the second, what is the mechanism of the coupling. The first question is relatively easy to answer.

sites of oxidative phosphorylation

How would you set about it?

One could look at the P/O ratios in submitochondrial complexes, or in fragments of the chain isolated by inhibitors or artificial electron donors or acceptors, or one could use the crossover theorem.

In fact, the first paragraph above already gives you some information about the probable location of one of the 3 phosphorylation sites. Can you deduce where it is?

If the P/O ratio from succinic acid is 2, and from $NADH_2$ is 3, and the electron transport chain is common at CoQ and beyond, it follows that two phosphorylations must occur beyond CoQ, and one in the $NADH_2$ dehydrogenase→non-haem Fe^{2+} part of the chain.

A combination of these techniques allows the identification of the three sites as shown in Figure 4.

As with the sequence of the electron transport chain there is no absolute certainty about these results, however; the data obtained vary with the technique used, and are not always concordant.

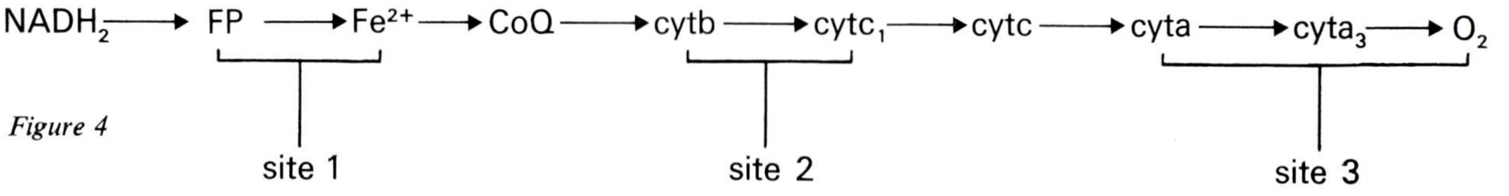

Figure 4

SAQ 19 (*Objective* 7)
List four reasons why results may disagree. (Answer on p. 41.)

One example of the sort of problem that may occur is that the site of inhibition by the barbiturate amytal is different in phosphorylating preparations, where it blocks between $NADH_2$ dehydrogenase and cyt b, and non-phosphorylating preparations, where it blocks between $NADH_2$ and the dehydrogenase.

You should now read the section in *Cell Structure and Function* headed Oxidative Phosphorylation (pp. 318–19). Note that the set book refers to α-oxoglutarate by the name that used to be conventional for the substance, α-ketoglutarate. In the α-oxoglutaric acid oxidation, which is an oxidative decarboxylation similar to that we considered in Section 3.5.3 for pyruvic acid, GTP (guanosine triphosphate) is formed first, and later converted to ATP. You need not worry about this variation.

In the next section in *Cell Structure and Function*, p. 319, on *coupled* phosphorylation, note the use of the term '*coupling*'. It was also dealt with briefly in *Chemistry of Life*, pp. 118–19, and in relationship to control mechanisms in the cell, discussed in *Chemistry of Life*, pp. 191–6, which you read in S100[29]. You should refresh your memory of the discussion there. Then read the next major section in *Cell Structure and Function*, pp. 320–3.

Note that *Cell Structure and Function* continues to use the 'high energy' formulation which we recommend discarding. Ignoring the terminology, their statement of the facts is in general acceptable.

On p. 322, note the use of a further technique for the investigation of mitochondrial ATP production: that of uncoupling agents, such as 2,4-dinitrophenol (DNP). Another agent used in the study of the phosphorylation mechanism is the inhibitor oligomycin, an antibiotic.

coupling and uncoupling

Note that in the reactions on p. 321 in *Cell Structure and Function*, if (3) and (4) are allowed to run in reverse, they function as an ATP-hydrolysing enzyme system, an ATP-ase. Uncoupling agents such as DNP, by dissociating reactions (1) and (2) from (3) and (4) will activate the ATP-ase. Inhibitors, on the other hand, may block individual reactions, such as (3), thus *inhibiting* the ATP-ase. Combinations of uncouplers and inhibitors may be used to unpick the sequence, as with the combination of inhibitors and electron acceptors and donors in the electron transport chain. The set book does not make this distinction between uncouplers and inhibitors very clear. However, it is worth emphasizing that little progress has been made towards identifying the hypothetical intermediates and only reactions (3) and (4) in the scheme in *Cell Structure and Function* are known to occur (although even with these the interpretation is not unequivocal) as a result of the type of isotope experiments discussed there. It is precisely because of the difficulty of isolating intermediates that the Mitchell scheme, known as *chemiosmosis*, has proved attractive to many. The Mitchell scheme depends on a recognition that the structure of the mitochondrion is crucial for its adequate functioning; ATP generation demands the hydrophobic (that is, water-excluding) environment of the lipoprotein of the cristae; in the presence of a hydrophilic (that is, water-containing) environment, hydrolysis (the ATP-ase reaction) occurs instead. It has been proposed by Racker (see *Cell Structure and Function*, pp. 324–5) that the phosphorylation mechanism is contained in the inner membrane spheres which can be seen in the cristae. Under appropriate conditions, the particles can indeed function as ATP-ase preparations. One explanation could be that the ATP-ase enzyme system is then activated (by uncouplers like DNP) by a conformational change in the protein structure so that the area associated with the ATP-generating reaction sequence is exposed to an aqueous environment, instead of, as in the phosphorylating system, water being rigorously excluded from it; in the aqueous environment, ATP hydrolysis rather than generation occurs.

ATP-ases

Once again, as with electron transport, but to an even greater degree, the experiments necessary to resolve the question of the coupling of oxidation and phosphorylation are complex and their interpretation difficult. Although the difference between the chemiosmotic and the 'high-energy intermediate' theories may seem slight in the terms in which *Cell Structure and Function* describes them, they have generated a reverberating scientific polemic, with experiments and counter-experiments echoing through the pages of *Nature*, the *Biochemical Journal*, *Journal of Biological Chemistry*, *Biochimica et Biophysica Acta* and innumerable symposia. We cannot explore the details of these controversies, but those of you going on to third-level biochemistry will have a chance to follow them in greater depth.

SAQ 20 (*Objective* 5)

Which is the best estimate of the overall percentage contribution of the mitochondria to a typical mammalian cell's ATP-generating capacity.

(a) 10 per cent.
(b) 30 per cent.
(c) 50 per cent.
(d) 70 per cent.
(e) 90 per cent.

SAQ 21 (*Objective* 10)

Which of the following is an example of:

(a) an uncoupling reaction;
(b) an inhibition reaction;
(c) neither.

(i) The effect of barbiturate on $NADH_2$ oxidation.
(ii) The effect of cyanide on cyt a_3 oxidation.
(iii) The effect of DNP on ATP hydrolysis.
(iv) The effect of oligomycin on ATP hydrolysis.
(v) The effect of removal of oxygen on cyt a_3 oxidation.
(vi) The effect of cyanide on ATP generation.

SAQ 22 (*Objective* 11)

During electron transport, which of the following sequences can involve the synthesis of ATP:

(i) NAD→CoQ.
(ii) Succinic acid→CoQ.
(iii) NAD→flavoprotein.
(iv) cyt c→cyt a.
(v) cyt a→O_2.
(vi) CoQ→cyt c.
(vii) CoQ→cyt b.

3.9 The Physiological Behaviour of Mitochondria

So far we have been concerned with the internal structure of the mitochondrion in relationship to function. But mitochondria do not normally exist in isolation from the cytoplasm; rather, they are constantly in interaction with it; responding to changes in cytoplasmic ATP, ADP and phosphate levels by changed rates of ATP synthesis; accepting hydrogen from cytoplasmic $NADH_2$ and passing out into the cytoplasm waste products of metabolism, such as water and CO_2. A good account of some of these functional interrelations is given in *Cell Structure and Function*, pp. 326–30, which you should now read. (See also Units 4 and 5.)

This Unit has been about energy generation in the cell, and therefore we have concentrated upon the energy-linked roles of the mitochondrion, analysing the organelle in terms of its machinery for oxidative phosphorylation. It would be wrong, however, to give the impression that this was all the mitochondrion was, and in recent years it is becoming increasingly recognized that the mitochondrion has a multi-functional role, containing a large variety of enzyme systems other than those we have yet considered. Thus all mitochondria contain enzyme systems for the oxidation of fatty acids, the synthesis of some amino acids such as glutamic and aspartic acids and the fixation of ammonia in urea (the 'urea cycle') although often similar enzyme systems are present in the cytoplasm as well. In fact mitochondria clearly have an indispensable part to play in a variety of cell functions. But the most intriguing observation of all has been that mitochondria contain their own DNA, RNA, ribosomes and protein synthetic capacity. This DNA is described in *Cell Structure and Function*, pp. 342–4, which you should now read. It is this combination of properties which has led to the proposal that the evolutionary origin of the mitochondria might have been a bacterial infection of an organism without mitochondria, followed by a subsequent symbiotic relationship.

mitochondrial function

DNA and RNA synthesis

SAQ 23 (*Objective* 13)

(a) Which of the following reactions can purified mitochondria *not* carry out:

(i) The oxidation of glucose.
(ii) The oxidation of pyruvic acid.
(iii) The synthesis of amino acids from carboxylic acids.
(iv) The synthesis of glucose.
(v) The synthesis of pyruvic acid.
(vi) The synthesis of RNA.
(vii) The synthesis of protein.
(viii) The uptake of Ca^{2+}.

(b) Which of the above reactions would you regard as of major functional significance in the mitochondrion *in vivo*.

3.10 The Chloroplast and the Fixation of Solar Energy

The ATP generating mechanisms we have considered are common to all living forms. They depend, as you know, on the breakdown of preformed organic substrates, notably glucose, and you know that the almost exclusive source of this is the photosynthetic mechanism of green plants. The mechanism of photosynthesis was very briefly discussed in S100[30], and those of you who have done S22– will have studied this more fully in Unit 3 of that course. Here we are concerned primarily with the analogies between mitochondrial oxidative phosphorylation and glycolysis, and the carbon-fixing reactions of the chloroplast. There is a good account of some of these in *Cell Structure and Function*, pp. 331–42 which you are now recommended to read. Note that for those of you doing S221, this material will be partly recapitulation. You are also advised that, if you are falling behind in your work in the Course you should omit the material on photosynthesis entirely. Although you will not be able to fulfil all the Unit objectives without following it, the account is self-contained. In reading the discussion in the set book, note that you need only recall the terms which appear in Table A: bear the Unit objectives in mind as you read.

photosynthesis

METABOLIC PATHWAYS CHART

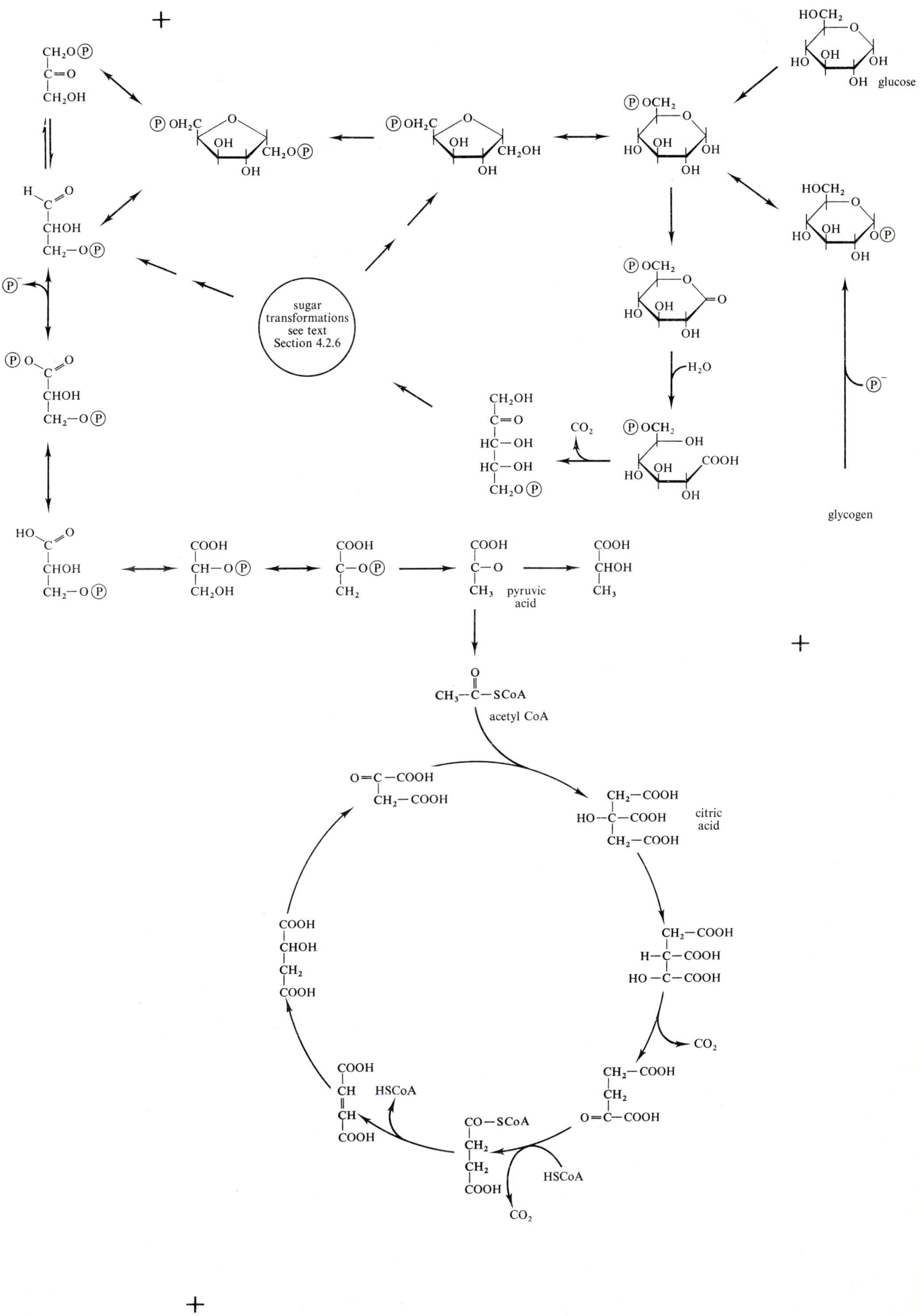

pp. 331–2 There are, in fact, other excitable molecules involved in photosynthesis, especially in bacteria.

p. 332 You should remember the general structure of chlorophyll but not its details. Note its resemblance to the cytochromes (Appendix 1). Note too that the nomenclature of chlorophyll is American; the British nomenclature is different but need not concern you here.

p. 333 Note the involvement of NADP rather than NAD in the photosynthetic reactions (Unit 4).

p. 334 The 'molecular lens' analogy should not be pressed too far. 'Robert' Hill should read Robin Hill. The Hill reaction refers to the ability of isolated chloroplasts to evolve oxygen in light provided they are supplied with an artificial electron acceptor. Note the analogies between the photosynthetic chain and the electron transport chain.

p. 335 You should recall the term 'photosynthetic phosphorylation' Phenazine methosulphate (note the English spelling!) is an artificial electron acceptor similar to the methylene blue and ferricyanide discussed in Section 3.7.4.
The calculation for 'efficiency of energy conversion' of 25 per cent should not be taken seriously, for reasons, discussed in Section 3.2, which should be familiar to you.

pp. 336–7 If you have it to hand, look also at the figures in *The Microstructure of Cells*,* pp. 45–8.

pp. 339–40 The distinction between light and dark reactions is important. You should read the account of the Calvin cycle, and note that many of the products were identified as a result of short-term isotope experiments of the sort discussed in TV 4. The account of the transformation of the various molecules which is given is rather complex to follow, and you should certainly not try to remember the names of any of the intermediates. What is interesting is the flow path of the carbon skeletons involved, and these are best understood by looking at Figure 5.

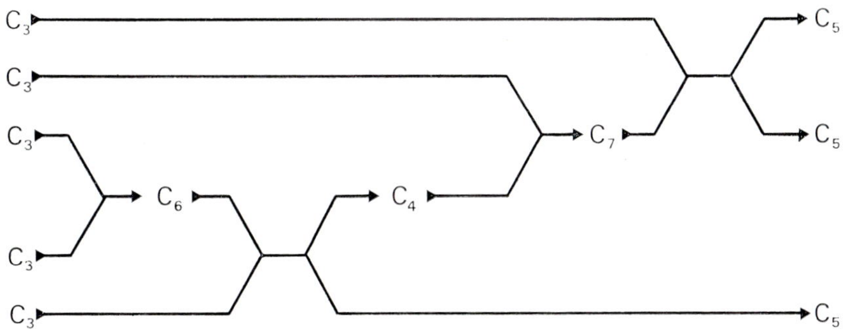

Figure 5

Note the general energy-balance of the reaction sequences, indicated on p. 341. The $NADPH_2$ and ATP are, of course, generated during the light reaction. While there are points of similarity between electron transport in the mitochondrion and energy transduction in the chloroplast, the one being in some senses the reverse of the other, the dark reactions of the Calvin cycle do not at first sight look at all like a reversal of the reactions of the citric acid cycle or glycolysis. As you will learn in Unit 4, the reversal of these reactions is far from straightforward. It is therefore of considerable interest to note that there *is* a pathway of glucose oxidation in mammalian systems which is very much more like a reversal of the Calvin cycle system. It is called the 'pentose phosphate pathway', which may carry, depending on the tissue, species and circumstances, some 15–30 per cent of the total glucose catabolism of the cell, the remainder, of course, going via the glycolytic (Embden-Meyerhof) route. The significance of the pentose phosphate pathway from the point of view of control and of evolution, and its relationship to photosynthesis, will also be discussed in Unit 4. For now, you may note that the relationships of all these pathways are spelt out in the diagram given in S100[31].

light and dark reactions

* *Stephen W. Hurry (1966, revised ed.)* The Microstructure of Cells, *John Murray.*

SAQ 24 (*Objective* 14)

Classify each of the items (a)–(c) according to the following key.

KEY A A false statement about photosynthesis.
B A true statement about photosynthesis.
C A statement about photosynthesis that may be true or false; it is not possible to decide.

(a) The dark reactions of photosynthesis only occur at times when the intensity of light falling on photosynthetic systems is very low.

(*b*) ATP and $NADPH_2$ are synthesized during photosynthetic phosphorylation.

(*c*) The membrane system of the chloroplasts is responsible for the light reaction of photosynthesis including the production of $NADPH_2$ and ATP.

(d) The first stable product of the dark reactions is a triose phosphate.

(e) A photosynthetic unit consists of 2 500 chlorophyll molecules, about 40 of which form the energy sinks.

(f) Quantasomes are the sites of the photosynthetic reactions.

(g) Electron flow energy is coupled to ATP synthesis by a mechanism similar to that found in mitochondrial-coupled phosphorylation.

(h) Carbohydrate is the end-product of photosynthesis.

3.11 Summary of Unit 3

In the first part of this Unit, we discussed critically the application of thermodynamic concepts to biochemical systems. In particular, we raised the question of the significance of the terminology 'high energy' when applied to substances like ATP. Instead the term 'group transfer molecules' was introduced. Some of these molecules and their roles were then reviewed, including ATP and creatine phosphate, and the sulphur-containing coenzyme CoA. The range of cellular activities involving ATP was emphasized.

Then, attention was turned to the reaction systems generating ATP, and the mechanisms of two, the glyceraldehyde-3-phosphate dehydrogenase and pyruvic decarboxylase reactions, discussed in some detail. Against this background, the organization of the principal ATP-generating reactions in the cell, those of the mitochondrion, were discussed. Because study of the mitochondrion is complex and highly dependent on experimental techniques, these were reviewed and the criteria for assessing a substance as an intermediate in the electron transport chain considered. A 'minimally agreed' formulation of the electron transport chain was presented. The coupling of oxidation with phosphorylation was considered, and the alternative theories of high-energy intermediate and chemiosmosis mentioned.

We then discussed the role of the mitochondrion in the cell, in terms of mitochondrial/cytoplasmic interactions and the multi-functional status of the mitochondrion. This led to speculations as to its evolutionary origin.

Finally, the role of the chloroplast in photosynthesis was related to that of the mitochondrion in oxidative phosphorylation, and the reactions by which glucose is synthesized during photosynthesis (the Calvin cycle) were reviewed in outline.

Appendix 1 Some Structural Formulae

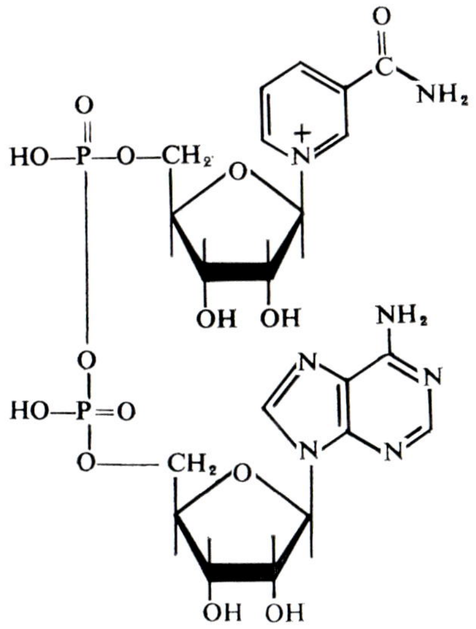

nicotinamide adenine dinucleotide (NAD)

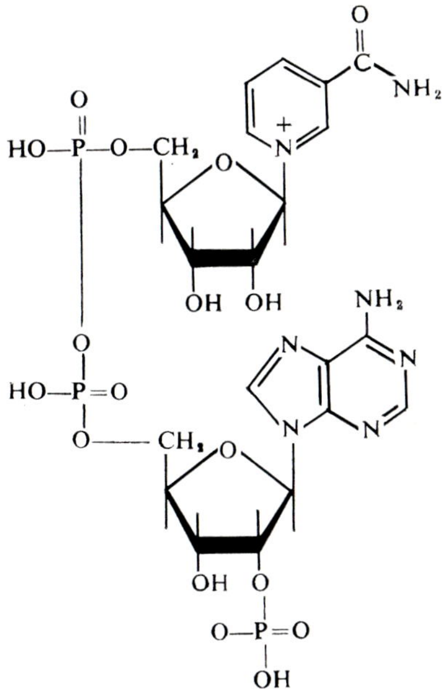

nicotinamide adenine dinucleotide phosphate (NADP)

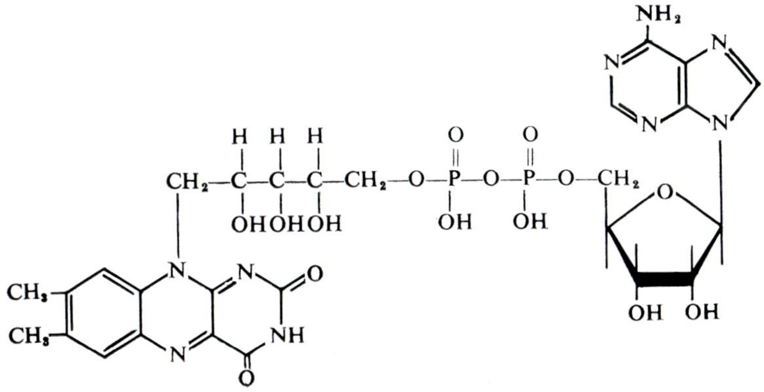

flavine adenine dinucleotide (FAD), the flavine prosthetic group of succinic dehydrogenase. The flavine prosthetic group of $NADH_2$ dehydrogenase is similar but not identical

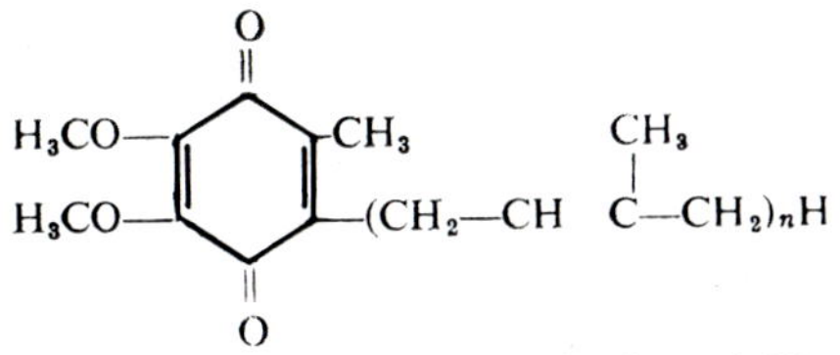

coenzyme Q (where *n* may be from 6–10)

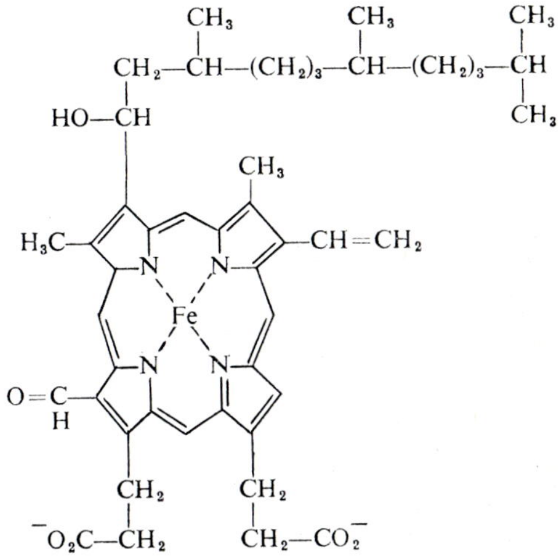

haem A, the prosthetic group of the cytochromes

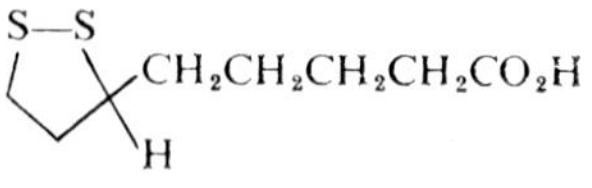

lipoic acid

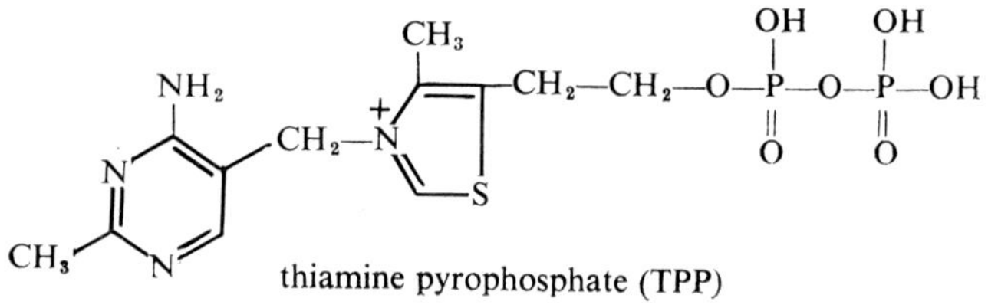

thiamine pyrophosphate (TPP)

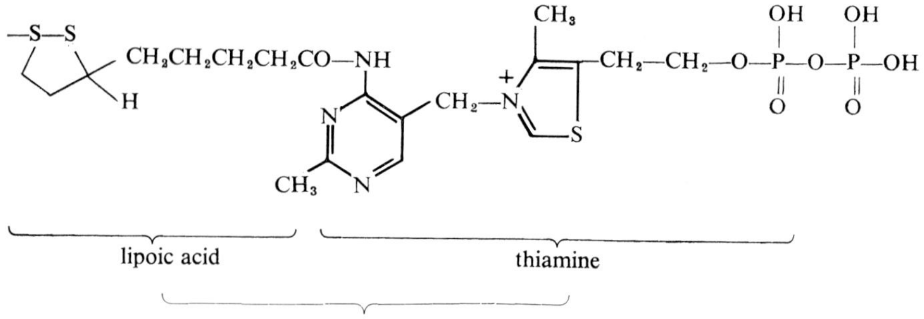

lipothiamide diphosphate

Further Reading

As well as the accounts of oxidative phosphorylation and cell energetics given in the standard textbooks listed in the Study Guide, the following book makes interesting reading: D. Keilin (1966), *The History of Cell Respiration and Cytochrome*, Cambridge University Press.

There is a good account of the mitochondrion and chloroplast from the points of view of several different researchers, in the collections of *Scientific American* readings: *Bio-organic Chemistry* (1968) and *The Living Cell* (1965), both published by Freeman.

See also A. L. Lehninger (1964) *The Mitochondrion*, Benjamin.

References to S100

1 Section 15.4, Unit 15, *Cell Dynamics and the Control of Cellular Activity.*
2 Section 11.1.4, Unit 11, *Chemical Reactions (1).*
3 Section 15.7.2, Unit 15.
4 Section 15.7, Unit 15.
5 Section 5.6.1, Unit 5, *The States of Matter.*
6 Sections 11.1.4; 11.2.3; 12.4.1, Unit 12, *Chemical Reactions (2).*
7 Section 12.2.1, Unit 12.
8 Section 5.2, Unit 5.
9 Section 5.2.1, Unit 5.
10 Section 15.4, Unit 15, and *Chemistry of Life*, pp. 82–5.
11 Section 15.6.3, Unit 15.
12 Sections 15.6.4 and 15.6.5, Unit 15.
13 Section 15.6.5, Unit 15, and *Chemistry of Life*, pp. 137–49.
14 Section 11.1.3, Unit 11.
15 Section 15.6.3, Unit 15, and *Chemistry of Life*, p. 113.
16 Sections 21.7 and 21.8, Unit 21, *Unity and Diversity.*
17 Section 21.8, Unit 21.
18 Section 16.1.3, Unit 16, *Cell Dynamics and the Control of Cellular Activity.*
19 Section 16.1.4, Unit 16.
20 Sections 17.8 and 17.9, Unit 17, *The Genetic Code: Growth and Replication.*
21 Section 16.3, Unit 16; Section 18.3.1, Unit 18, *Cells and Organisms.*
22 Section 16.3, Unit 16; Section 18.3, Unit 18.
23 Section 15.6.1, Unit 15.
24 Section 16.4, Figure 2, Unit 16.
25 TV programme for Unit 14.
26 Section 15.7.1, Unit 15.
27 Radio programme for Unit 15.
28 Section 15.6.2, Unit 15.
29 Section 16.3, Unit 16.
30 Section 15.8, Unit 15.
31 Section 15.5, Figure 7, Unit 15.

Self-assessment Answers and Comments

SAQ 1 1 (a); 2 (a); 3 (a); 4 (b); 5 (a); 6 (b); 7 (a); 8 (b); 9 (a); 10 (c).

SAQ 2 In general, in the groups capable of participating in phosphoryl transfer, the phosphate group is close either to another phosphate or to a resonating organic radical capable of entering into conjugation with the mobile electrons of the phosphate group. By contrast, the phosphate groups which cannot transfer, as in diphosphoglyceric acid, may be represented R—O—P where R is a saturated organic radical or one in which the conjugated fragment is separated from the phosphate.

SAQ 3 (d).

SAQ 4 (e) and (f).

SAQ 5 (c), (d) and (g).

SAQ 6 (a), (b), (c), (e) and (f).

SAQ 7 (a).

SAQ 8 (d) is the best answer; (a), (b) and (e) are factually incorrect; (c) is correct but incomplete; (f) is correct but many other reactions also generate $NADH_2$ from NAD.

SAQ 9 nucleus > mitochondrion > lysosome > ribosome.

SAQ 10 Two; an inner and an outer.

SAQ 11 The second statement is correct.

SAQ 12 All of the criteria except the first one are of relevance (it doesn't matter if the substance is present in excess; only that it is there in the right quantity). However, not all are necessary and sufficient. Thus (4) and (5) are specific to particular chain components but will not apply to all. (6) and (7) are each partial criteria. That is, they are conditions which a component of the chain should fulfil, but they are not in themselves both necessary or sufficient; the combined criterion (8) is a necessary one and a better statement than either (6) or (7) alone. (2) and (3) are both necessary. (2), (3) and (8) are together both necessary and sufficient.

SAQ 13 *Figure* 6

SAQ 14 *Figure* 7

SAQ 15 (b); the others are incorrect.

SAQ 16 (c); the others are incorrect.

SAQ 17 (f) is the best statement; (a) and (b) are partial statements; (c), (d), (e) and (g) are false.

SAQ 18 $cyta_3$: (d), (h), (j).
FAD: (b), (g).
NAD: (c), (g).
CoQ: (e), (g)
Vit K: (e), (f).
SDH: (a), (g).
NADP: (c), (f), (g).

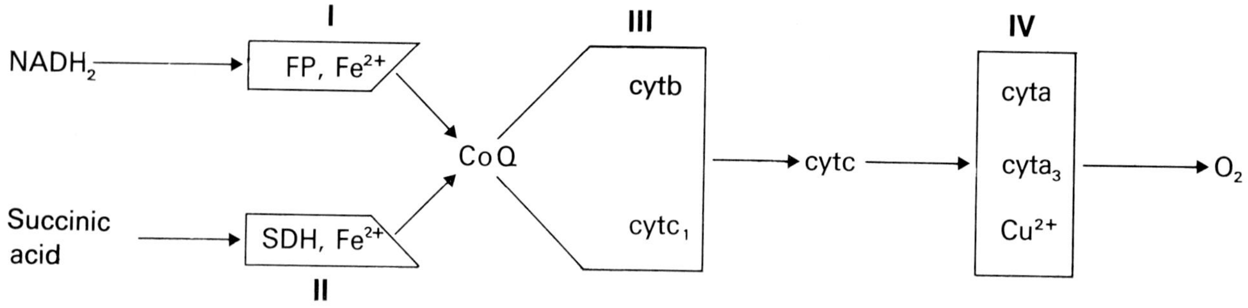

Figure 6

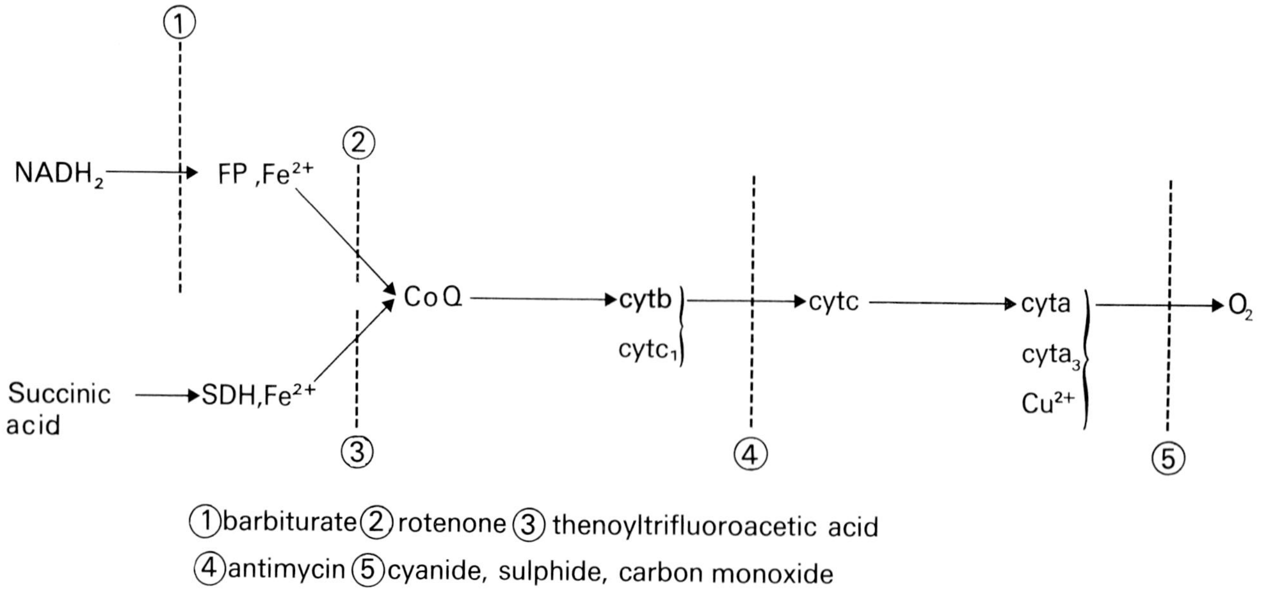

Figure 7

SAQ 19 We would include the following:

1 Submitochondrial fractions might be damaged.

2 The fractions might be cross-contaminated and impure.

3 An inhibitor may block more than one site.

4 An electron donor or acceptor may be involved at more than one site though not always with the same degree of efficiency. But you may think of more.

SAQ 20 (d).

SAQ 21 (i) (b).

(ii) (b).

(iii) (a).

(iv) (b).

(v) (c) – this is not inhibition as we have used the term probably.

(vi) (c) – although cyanide will stop ATP production it acts indirectly by blocking the reoxidation of cyt a_3. But according to some recent evidence it might also affect coupling at Site 3.

SAQ 22 (i); (v); (vi).

SAQ 23 (a) (i) and (iv). (v) can occur by way of the citric acid cycle and CO_2 fixation, though we do not consider the significance of this reaction until Unit 4.

(b) This is a harder one, and an element of judgement is involved. We would say certainly (ii), probably (vi) and (vii), and perhaps (viii). But future research may prove us wrong!

SAQ 24

(a) A false statement – see *Cell Structure and Function*, pp. 338–9.

(b) A false statement – see *Cell Structure and Function*, p. 335 and Figure 13.9, p. 333; only ATP is produced by this process.

(c) This statement is made on p. 341 and contrasts with those on p. 338. It is a weak generalization at best and there are no grounds for extending it from chloroplasts in green plants to all chloroplasts or to the membranes of non-green plants and bacteria involved in photosynthesis.

(d) A true statement – see *Cell Structure and Function*, p. 339. The stability is implied by the radioactive labelling experiments.

(e) The statement is inferred from experiments, which are not cited in *Cell Structure and Function*, p. 338. The figures for the numbers of the two kinds of molecule are estimates only.

(f) A false statement, as quantasomes are substructure of the grana membranes while photosynthesis involves membrane reactions and 'soma reactions'.

(g) False—see *Cell Structure and Function*, p. 335.

(h) False—see *Cell Structure and Function*, p. 341.

Contents

* This Chart is inserted into the Unit as loose sheets.

Table A

List of Scientific Terms, Concepts and Principles used in Unit 4

Taken as prerequisites			**Introduced in this Unit**			
1 Assumed from general knowledge	**2 Introduced in a previous Unit**	**Unit No.**	**3 Developed in this Unit**	**Page No.**	**4 Developed in a later Unit**	**Unit No.**
		S100*				
	metabolic pathway	15, 16	central pathway	7	allostery	5-6
	anabolism	15, 16	activation reaction	8		
	catabolism	15, 16	phosphofructo-			
	glycolysis	15, 16	kinase	9		
	citric acid cycle		gluconeogenesis	10		
	(Krebs cycle)	15, 16	hexokinase	10		
	pyruvic acid	15, 16	branching point	10		
	lactic acid	15, 16	glucose-6-phosphate	10		
	glucose	15, 16	fructose-1,6-			
	sugar phosphates	15, 16	diphosphatase	11		
	phosphorylation	15, 16	pentose phosphate			
	ATP	15, 16	pathway	12		
	ADP	15, 16	triose	12		
	AMP	15, 16	tetrose	12		
	oxidation-		pentose	12		
	reduction		hexose	12		
	reactions	15, 16	heptose	12		
	glycogen	15, 16	transketolase	12		
	cytoplasm	14	transaldolase	13		
	mitochondrion	14	C–1/C–6 ratio	14		
	isomerization	10	mass action	16		
	hydrolysis	13	Pasteur effect	16		
	photosynthesis	15, 16	adipose tissue	18, 29		
	hormones	18	lipolysis	18		
	adrenalin	18	fatty acid	18		
	thyroxin	18	glycerol	18		
	stoicheiometry	10	triglyceride	18		
	ester	10	alanine	20		
			glutamic acid	20		
		S24–**	aspartic acid	20		
	acyl group	8	asparagine	20		

* *The Open University (1971) S100* Science: A Foundation Course, *The Open University Press.*

** *The Open University* (1972) *S24–* Science: A Second Level Course, An Introduction to the Chemistry of Carbon Compounds, *The Open University Press.*

Taken as prerequisites			**Introduced in this Unit**			
1 Assumed from general knowledge	**2 Introduced in a previous Unit**	**Unit No.**	**3 Developed in this Unit**	**Page No.**	**4 Developed in a later Unit**	**Unit No.**
		S2–1	oxidative deamination	21		
	coenzyme A (CoA)	3	glutamic dehydrogenase	21		
	acetyl CoA	3	glutamic transaminase	21		
	pyruvic dehydrogenase	3	urea	21		
	isocitric dehydrogenase	3	anaplerotic pathway	22		
	vitamins	2	glyoxylic acid sequence	22		
	transamination	2	pyruvic carboxylase	22		
			ketosis	23		
			diabetes	24		
			insulin	24		
			mitochondrial permeability	24		
			acyl carrier protein	25		
			acetyl CoA carboxylase	25		
			malonyl CoA	25		
			glycogenolysis	29		
			glucagon	31		
			phosphorylase	31		
			phosphatase	32		
			kinase I	32		
			kinase II	32		
			adenyl cyclase	32		
			cyclic AMP	32		
			diester	33		
			second messenger	33		
			enzyme cascade	33		
			glycogen synthetase	33		
			β-oxidation of fatty acids (*ITQ* 2)	38		

Objectives

When you have completed this Unit, you should be able to:

1 Demonstrate your knowledge of the items in Table A by being able to define them in your own words, or recognize the best definition of them, or distinguish between correct and incorrect statements concerning them.

2 List the major anabolic, catabolic, and central pathways, and describe their important features. (*SAQ*s *1, 2, 5, 7.*)

3 Name or, when presented with a reaction sequence, identify the major established control points of carbohydrate metabolism.

4 Show understanding of the experimental basis of the study of metabolism by:

(a) when given an hypothesis concerning the pathway of breakdown or synthesis of a metabolite, or the control of such a pathway, devising an experiment to test it or to distinguish alternative hypotheses concerning the metabolite or the pathway;

(b) when presented with experimental data concerning metabolic processes, distinguishing between warranted, unwarranted and contradicted conclusions drawn from them. (*ITQ*s *1–8*, *SAQ 9.*)

5 Given the activation and inhibition characteristics of a particular enzyme or enzymes in a metabolic pathway, deduce the effects of various constraints on the rate of reaction of particular steps of the pathway, and on the levels of particular intermediates of the pathway. (*SAQ 6.*)

6 Given a pathway or important reaction, name: (a) in which cell compartment it occurs; (b) in which tissues of the mammalian organism it occurs; (c) in which types of organism it occurs. (*SAQ*s *4, 7.*)

7 If presented with a diagram of the glycolytic pathway, recognize the points at which the reverse pathway, gluconeogenesis, differs from it, and be able to comment on the significance of these differences.

8 Describe the major features of regulatory enzymes, and comment on the significance of these features. (*SAQ 3.*)

9 Show comprehension of the roles of the various metabolic pathways by predicting the behaviour of systems in which particular pathways do not occur, or are inhibited. (*SAQ 8.*)

Study Guide

There is no reference out to your set book, *Cell Structure and Function*,* in the text of this Unit, as the approach adopted there is too different from ours. Before starting on the main text of the Unit, you should do the Pre-Unit Assessment Test. This tests the knowledge from S100, Units 15 and 16, that we assume. If you have difficulty with this, scoring less than two-thirds correct, you should revise the material by reading S100, Unit 15, Sections 15.5, 15.6.3, 15.6.4, 15.6.5, 15.7.3 and Unit 16, Sections 16.1, 16.2 and 16.3, including the passages from *The Chemistry of Life*.** For a more continuous account, with a rather different approach, read instead the whole of Chapter 12 of *Cell Structure and Function*, without attempting to learn any material which is new to you.

If you find yourself very short of time, Sections 4.8.2, 4.8.3 and 4.8.4 may be omitted, as nothing later on in the Course hinges on these Sections. However, the concepts included there are quite important ones for a balanced view of the subject, so don't leave them out unless you have to!

Accompanying the Unit is a Metabolic Pathways Chart (described in more detail in Section 4.1), which you will have to refer to at several points in your reading of this Unit.

* *A. G. Loewy and P. Siekevitz (1969)* Cell Structure and Function *(2nd ed.), Holt, Rinehart and Winston.*

** *Steven Rose (1970)* The Chemistry of Life, *Penguin Books.*

Pre-Unit Assessment Test (for recall of S100)

1 Anabolism and catabolism are two of the types of metabolic pathway dealt with in this Unit. In each of the following pairs, one term applies most frequently to anabolism, the other most frequently to catabolism. Which is which?

1	Synthesis	Breakdown
2	Linked to ATP production from ADP and phosphate	Linked to ATP breakdown
3	Energy-using	Energy-yielding
4	Linked to $NADH_2$ production from NAD	Linked to NAD production from $NADH_2$

2 (i) The glycolytic pathway, under aerobic conditions, breaks down glucose to which of the following:

(a) pyruvic acid
(b) ethanol
(c) citric acid
(d) lactic acid
(e) carbon dioxide + water

(ii) Which of (a)–(e) above is produced under anaerobic conditions by yeast?

(iii) Which of (a)–(e) above is produced under anaerobic conditions by mammalian muscle?

3 What is the name of the series of reactions by which pyruvic acid is broken down to carbon dioxide and water?

4 Name two biochemical reactions linked with the catabolism of glucose to pyruvic acid.

5 Match each of the biochemical pathways or reactions (list A below) with the subcellular fraction with which it is most clearly associated (list B):

List A	*List B*
Glycolysis	Ribosomes
Protein synthesis	Mitochondria
Oxidative phosphorylation	Nucleus
Citric acid cycle	Endoplasmic reticulum
	Golgi bodies
	Cytoplasm

For answers see p. 37.

4.1 Metabolic Pathways

Study Comment

The description of the pathways in this Unit relies heavily on the use of the Metabolic Pathways Chart (which is enclosed with this Unit). This details glycolysis, the citric acid cycle, gluconeogenesis, and deals briefly with another pathway of glucose breakdown, the pentose phosphate pathway. The main figure in each case (in black) shows the structural formulae only; the overlays name the intermediates and enzymes involved in the reactions under consideration. There are three overlays. Overlay A deals with the reactions involving ATP; overlay B shows the oxidation/reduction reactions; overlay C names the enzymes and metabolites which are discussed in detail in the text. When using this Chart, you should bear in mind that you are not expected to memorize these pathways, but you should be able to describe the reactions that are dealt with in detail in the text (*Objectives 2, 3, 6, 7*).

Up to now, the Course has been concerned mainly with various aspects of cells – organelles, enzymes, single reactions. This Unit is concerned with *metabolic pathways*, the coherent sequences of reactions that result in the breakdown of food, or of cell components, or in the synthesis of new cellular materials.

Three kinds of metabolic pathway are usually identified in an organism. You are familiar with catabolic (breakdown) and anabolic (synthetic) routes from S100[1].*

In this Unit another kind will be distinguished, the *central pathways*. These are the pathways involving the interconversions of substances that are both the products of catabolism, and the precursors of anabolism. The citric acid cycle may be considered as belonging to this group.

central pathways

The central pathways may thus be involved in either breakdown or synthesis, depending on the state of the cell, and the reactions involved form a complex network. The catabolic and anabolic pathways, on the other hand, form something much closer to a linear progression (although diagrams often exaggerate this. The diagrams in this Unit are no exception!)

Conventions and abbreviations

It is necessary at this point to introduce certain conventions and abbreviations which are used throughout this Unit.

1 The glucose molecule will be represented in its ring form, which is the form that normally occurs in solution. However, as you know, in glycolysis there is a step that involves the splitting of the glucose molecule into two three-carbon sugars, which are straight chain molecules, so Figure 1 is included to remind you of the equivalence of the two ways of representing glucose.

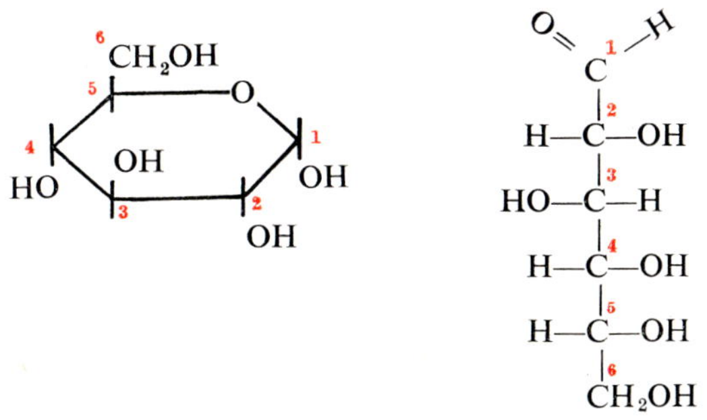

Figure 1

2 In normal chemical usage, the arrow → in a chemical equation means that the reaction is irreversible. A reversible reaction is shown by the symbol ⇌.

In the reaction sequences here, a different convention will be used. The single arrow → will be used to show the direction of a reaction *in vivo*. This may or may not mean that the reaction in question is thermodynamically irreversible. Where a given step may occur in either direction *in vivo*, it is shown by ⟷.

* *The Open University (1971) S100* Science: A Foundation Course, *The Open University Press. Superscript numbers indicate more detailed references to S100, which are listed on p. 40.*

4.2 Glucose Catabolism and Anabolism

This Section deals with the pathway by which glucose (or glycogen) is broken down to pryuvic acid, lactic acid or ethanol, and the pathway carrying out the reverse of this process. These two pathways have been chosen because they illustrate rather clearly some general features of metabolism. An alternative pathway of glucose breakdown is also dealt with.

The catabolism of glucose, of which glycolysis is the first stage, is one of the major sources of energy in almost all cells – in fact, some cells are totally dependent on glucose, e.g. adult mammalian brain cells. Glycolysis is thus of almost universal importance, and the main reaction sequence also seems to be essentially identical in all systems studied.

The net reaction of glycolysis is the conversion of glucose to pyruvic acid.

Two other reactions are linked to this conversion (S100[2]). What are they?

2ADP→2ATP
2NAD→$2NADH_2$

These reactions are linked to glycolysis, i.e. the sequence glucose→pyruvic acid cannot occur without the presence of ADP and oxidized NAD in the cell. The overall reaction may thus be written

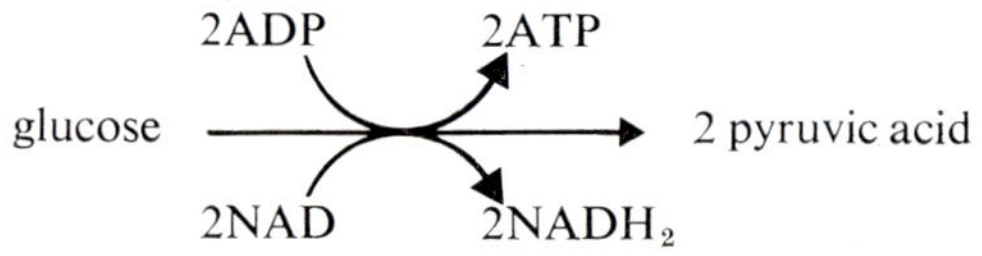

If the reaction is to continue, therefore, the ATP and $NADH_2$ must, in the steady state, be converted back to ADP and NAD at the same rate as they are produced. These conversions were outlined in Unit 3.

What kinds of reactions may involve the conversion of ATP→ADP

Synthesis of cell components.
Performance of mechanical work (muscle).

What reactions may convert $NADH_2$→NAD in animals?

The electron transport chain.
Conversion of pyruvic acid→lactic acid.

Which of these occurs under anaerobic conditions in animals?

Conversion of pyruvic acid→lactic acid.

4.2.1 The glycolytic sequence

The complete glycolytic sequence, including breakdown and synthesis of glycogen, is shown in the Chart. Put on overlay A, which shows the reactions involving ATP and ADP. Note that the reaction sequence begins with two phosphorylation reactions. Although the overall pathway of glycolysis results in a net synthesis of ATP from ADP, the initial stages involve use of ATP to 'activate' the glucose – sugar phosphates are much more reactive compounds than the unsubstituted sugars. Such *activation reactions* are frequently found in metabolic pathways. Another example, which will be seen later, is that of fatty acid breakdown, where the acid must first react with coenzyme A, a reaction linked to ATP breakdown. Later reactions in the glycolytic pathway, of course, are linked to ATP synthesis by the substrate level phosphorylation described in Unit 3, giving the net yield of ATP referred to above.

activation reactions

Now remove overlay A, and put on overlay B. This shows the oxidation/reduction reactions of the sequence, which you should be familiar with from your knowledge of S100[2] and from what has been said above.

4.2.2 Control points

Another common feature of metabolic pathways is that an enzyme catalysing an early reaction, often the first reaction specific to that pathway, frequently acts as a control point. In the case of glycolysis, *phosphofructokinase* (PFK) (see overlay C) is the main regulating enzyme.

control point

Which of the following properties would you expect to be possessed by a regulating enzyme?

1 It must have a very high activity in the tissue or organism.
2 It would be expected to have a rather low activity in the tissue organism.
3 It would be expected to have a high affinity for its substrate.
4 It must have a low affinity for its substrate.
5 It must be subject to activation or inhibition by metabolites in the cell.
6 It must be subject to competitive inhibition by substances similar to its normal substrate.

---

2 and 5 are necessary. The enzyme must be a rate-limiting step in the pathway if it is to regulate the whole pathway, and the regulation must be by normal cell components (at levels which normally occur). 1 is contradicted by the data (2 being correct); 3, 4 and 6 are irrelevant.

The enzyme, of course, need not be rate-limiting *all* the time, so it is not sufficient to determine a set of enzyme activities for the tissue, and then to say that the enzyme with the lowest activity is a regulating enzyme; in any case, some of the enzymes may be damaged by the extraction. What can be said, though, is that the activity of a regulating enzyme would be expected to be on the low side compared to other enzyme activities. Look at Table 1, which shows such a set of activities for some enzymes of carbohydrate metabolism.

Table 1 Maximal catalytic capacities of glycolytic enzymes of rat tissues

	μmol substrate converted/ g wet weight/min at 37°	
	Liver	Muscle
phosphofructokinase	3.3	80
pyruvic kinase	50	780
fructose diphosphate aldolase	10	77
glyceraldehyde-3-phosphate dehydrogenase	170	590
enolase	17	340

So, clearly, the first requirement for a regulating enzyme is met by phosphofructokinase.

4.2.3 Phosphofructokinase

Phosphofructokinase is also an *allosteric* enzyme. Unit 5 will discuss this subject further. It is sufficient to say here that an allosteric enzyme is inhibited or activated by specific metabolites that bear no particular structural relationship to the substrates. Studies on such enzymes lead to the conclusion that these activators or inhibitors act at different sites from the substrates.

allosteric enzyme

If phosphofructokinase is to act in regulating the level of glycolysis to the needs of the cell, which substances would you expect to activate it, and which to inhibit it?

The *products* of glycolysis would be expected to inhibit, e.g. pyruvic acid (or acetyl CoA), ATP, $NADH_2$. In fact, ATP is a strong inhibitor of the enzyme. The other metabolite which inhibits it is citric acid, which can also be considered a product of glycolysis, as it is produced further along the pathway. The *starting metabolites* of glycolysis would be expected to activate it, e.g. glucose, hexose* phosphates, ADP, or NAD. In fact, both ADP and AMP activate the enzyme.

---

* *Defined in Section 4.2.6.*

It has been said that an early reaction in a metabolic sequence is often a regulation point, PFK catalysing an early reaction specific to glycolysis. Looking at the Chart with overlay C in place, suggest two other enzymes involved in carbohydrate metabolism that might be regulatory.

---

1 Glycogen phosphorylase, catalysing the first reaction in glycogen metabolism. This enzyme is affected by hormonal mechanisms, and is subject to a different level of control. It will be dealt with later (in Section 4.8.3).

2 Hexokinase, catalysing the first reaction in glucose metabolism.

4.2.4 Hexokinase

This enzyme is important as the starting point of glycolysis, the pentose phosphate pathway (see Section 4.2.6) and of glycogen synthesis from glucose. There is very little free glucose within most cells, as it is phosphorylated very readily. This is related to the high affinity of hexokinase for glucose.

Hexokinase is an allosteric enzyme but, unlike phosphofructokinase, it is inhibited by its *immediate* product, glucose-6-phosphate, not by a metabolite further down the pathway. Hexokinase comes immediately before a *branching point* in glucose metabolism, while phosphofructokinase follows the branching point, at the beginning of a single pathway. **branching point**

Comment
The interpretations made here and later about regulation of enzymes need to be treated with a certain distrust. You have been introduced to teleology (S100[3]) and to the ways in which preconceived ideas, in this case concerning purpose, can influence the interpretation of experimental data. There is a more subtle kind of bias that can appear not in the interpretation of observations, but in the choice of which observations to make. Studying enzymes to find out which are regulating enzymes, and which compounds activate or inhibit them, is a time-consuming task, and so inevitably the research that is done is biased towards those enzymes that are known to be allosteric, or are suspected of being so. Similarly when testing compounds for their effects on an enzyme, the researcher will tend to choose compounds that he expects to have an effect. Thus the researchers' preconceived ideas may under certain conditions tend to be supported, i.e. when there is no simple 'critical experiment'. Are all enzymes regulating enzymes? Probably not, but the scheme presented here, even if broadly correct, is almost certain to be considerably 'neater' than reality.

4.2.5 The reversal of glycolysis

Another feature of metabolic pathways in general, and of glycolysis in particular, is that where a compound can be either broken down or synthesized, synthesis is never the exact reverse of breakdown. Parts of the pathway may occur by a simple reversal of enzyme reaction, but never the whole pathway. In most tissues, glycolysis can be reversed to produce glucose-6-phosphate, or glycogen, from pyruvic acid. Some tissues can also dephosphorylate the glucose-6-phosphate, and release free glucose into the bloodstream; mammalian liver and kidney cortex, for example. The production of free glucose from pyruvic acid is called *gluconeogenesis;* it is of considerable significance in fasting, when blood glucose is maintained largely at the expense of protein breakdown (Section 4.5). The reactions involved specifically in these reactions, the synthesis of glucose, glucose-6-phosphate, or glycogen are shown on the Chart. There are several steps that are different in the catabolic and anabolic sequences. Put both overlays A and C on to the Chart. There are four steps on the pathways from glucose or glycogen in which the reverse reaction has a different mechanism. **gluconeogenesis**

What is different about the stoicheiometry of the anabolic and catabolic reactions in *every case*? **stoicheiometry**

The number of molecules of ATP broken down in the anabolic reaction is greater than the number synthesized in the catabolic reaction.

Take, for example, phosphofructokinase and fructose-1,6-diphosphatase:

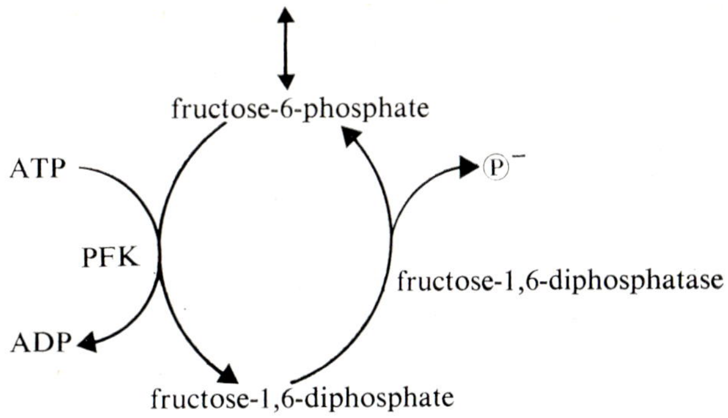

So, while the whole sequence of glycolysis has the stoicheiometry

gluconeogenesis is represented as

more ATP being required to synthesize glucose than can be obtained from its breakdown.

This is also a general feature of catabolic and anabolic routes. Because of the difference in the coupling to ATP broken down or synthesized, the equilibrium position of the synthetic sequence is different from that of the catabolic sequence. Both the overall reactions shown above are virtually irreversible. Which one occurs depends upon the activity of the enzymes unique to each pathway. It is not surprising, then, to find that, just as the enzymes controlling glycolysis tend to be those unique to glycolysis, so the enzymes regulating gluconeogenesis and glycogen synthesis are the corresponding enzymes shown on overlay C. These are pyruvic carboxylase, which is activated by acetyl CoA; fructose-1,6-diphosphatase, which is strongly inhibited by AMP; and glycogen synthetase, whose regulation is complex, and subject to hormonal control – it will be dealt with in Section 4.8.3 together with glycogen phosphorylase.

There is another feature of this control. In each of the cases where the anabolic and catabolic reactions differ, there is the possibility of the two reactions forming a cycle, e.g.

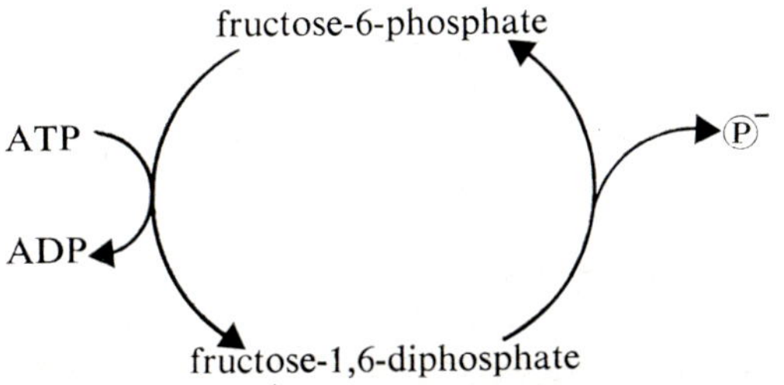

What will be the net result of this cycle?

The hydrolysis of ATP to ADP and phosphate.

Similar reactions are theoretically possible for each of the pairs of reactions considered. They are called *futile cycles* because they achieve no overall reaction apart from the hydrolysis of ATP. Clearly if ATP is not to be 'wasted' in this way, these cycles must be suppressed. The fact that both enzymes concerned are regulating enzymes makes this possible.

futile cycles

Glycolysis, as you know from S100[4], occurs in the cell cytoplasm. In which cell compartment would you expect the synthesis of glucose-6-phosphate from pyruvic acid to occur?

In the cytoplasm also, as many of the enzymes are in common between the two pathways.

4.2.6 The pentose phosphate pathway

sugar nomenclature

Note Before dealing with this pathway you should be reminded of the nomenclature for sugars with different numbers of carbon atoms. All the sugars you have met in any detail so far have had six carbon atoms. These are called hexoses. Sugars, however, can have different numbers of carbon atoms; these are named as follows:

3 – triose	5 – pentose	7 – heptose
4 – tetrose	6 – hexose	

These are general names, and there are many different sugars of each chain length, e.g. fructose, glucose and galactose are *all* hexoses; ribose, which occurs in RNA, and ribulose are both pentoses. You need not remember the names of the sugars in the reaction sequence that follows, but you should be able to reproduce a schematic diagram showing the number of carbon atoms in the sugars concerned.

Another pathway of glucose catabolism is also important, although usually breaking down less than half of the amount of glucose broken down by glycolysis. This pathway is called the pentose phosphate pathway because the first part of it consists of the production of a pentose phosphate according to the stoicheiometry:

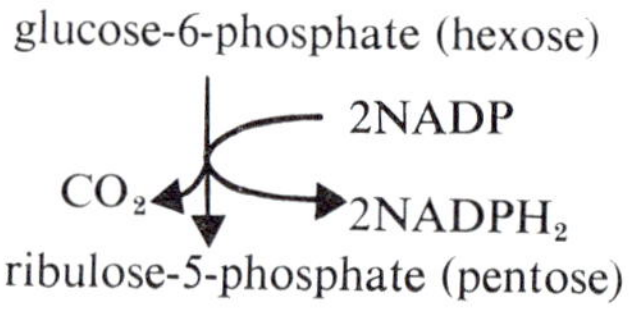

The details are shown on the Chart with overlay B, leading from glucose-6-phosphate to the circle labelled sugar transformations.

These sugar transformations, the second part of the reaction sequence, cannot be represented as a single pathway, but the overall reaction is:

3 pentose → 2 hexose + 1 triose

or, alternatively, because two triose molecules can combine to form one hexose molecule:

6 pentose → 5 hexose.

$NADPH_2$

The hexose can enter the pathway once more, and so this pathway can theoretically carry out the complete breakdown of glucose to $CO_2 + H_2O$ with the production of $NADPH_2$ (N.B. *not* $NADH_2$).

transketolase

The sugar transformations, which carry out this overall reaction, 3 pentose → 2 hexose + 1 triose, are based on two enzymes. Both of these are capable of transferring a part of one sugar molecule to another. Transketolase transfers a 2-carbon unit in the following way (note that the stereochemistry of the molecules has been ignored in these diagrams):

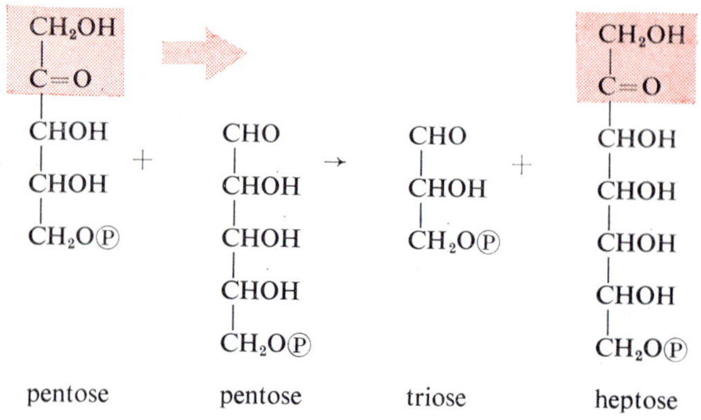

This is the first transfer reaction of the sequence. Transaldolase, on the other hand, transfers a 3-carbon unit. In the second transfer reaction of the sequences, transaldolase acts on the two sugars which are the products of the first reaction:

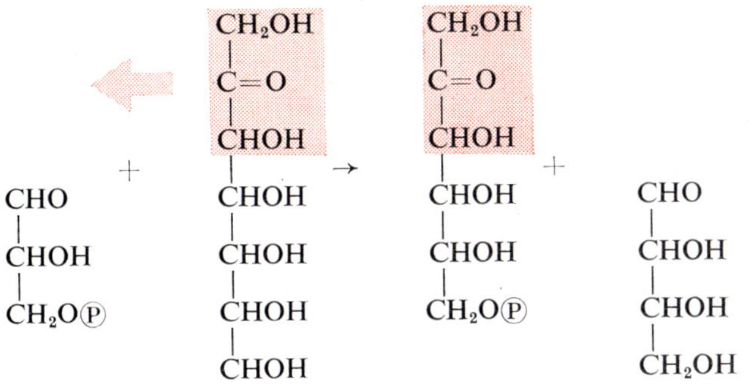

triose heptose hexose tetrose

transaldolase

You do not need to remember which enzyme does which, or the exact reaction. You should, however, be aware that one enzyme transfers a 2-carbon fragment, and the other a 3-carbon fragment.

The enzymes involved are not very specific about the length of the carbon chain of the sugar they act upon; in the next reaction transketolase acts upon the tetrose and a third pentose molecule,

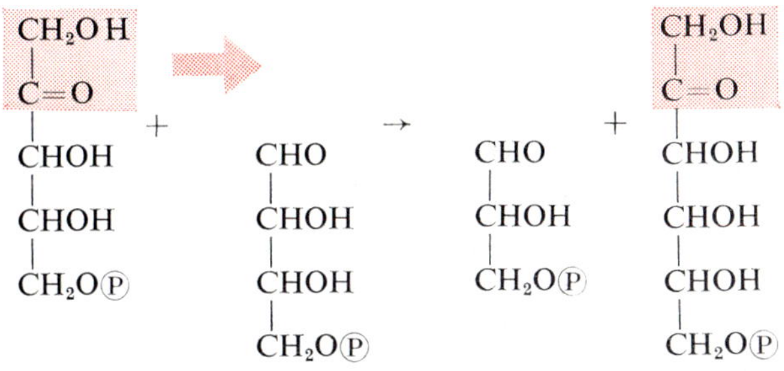

pentose tetrose triose hexose

transketolase

completing the process. This can be outlined in another way, by means of a flow diagram:

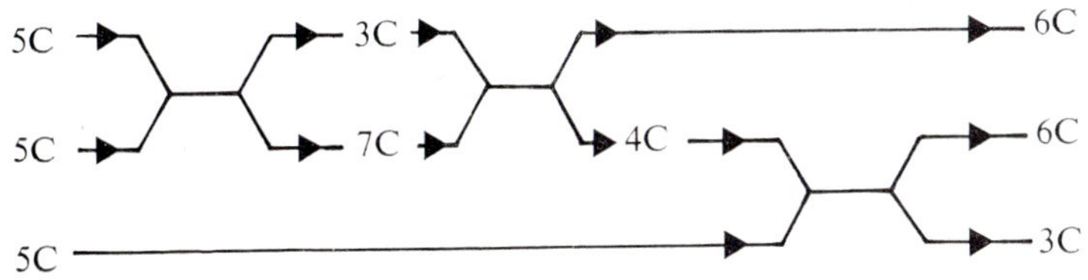

These reactions are rather similar to the *reverse* of the reactions that follow carbon dioxide fixation in the dark reactions of photosynthesis (Unit 3). The question now arises of how to measure the relative contributions of the two sequences to the total tissue breakdown of glucose.

It is possible to prepare radioisotope (^{14}C) labelled glucose in which only one specific carbon atom in the molecule is labelled. Consider four different types of glucose molecule, labelled in the 1, 3, 4 or 6 positions respectively:

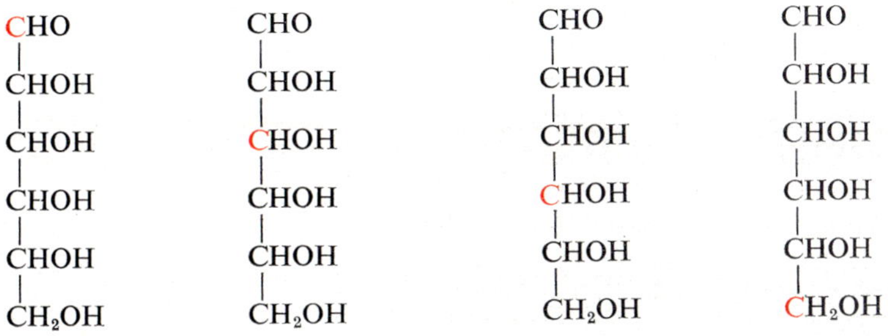

1-(^{14}C)-glucose 3-(^{14}C)-glucose 4-(^{14}C)-glucose 6-(^{14}C)-glucose

Refer to the Chart for the glycolytic sequence. Note that when the fructose-1,6-diphosphate is split into two triose phosphate molecules, glyceraldehyde-3-phosphate and dihydroxyacetone-phosphate, and the latter is then isomerized

(S100[5]) into a second molecule of glyceraldehyde-3-phosphate, the reaction is as shown in Figure 2.

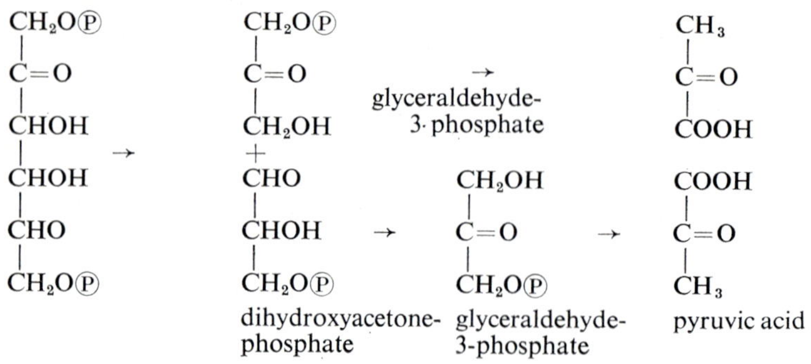

Figure 2

Note also that the pyruvic acid is then decarboxylated to acetic acid in the form of acetyl CoA. This reaction has been dealt with in some detail in Unit 3. The acetic acid can then be completely broken down to carbon dioxide and water.

> Name the pathway by which acetyl CoA is broken down to carbon dioxide and water.
>
> ---
>
> The citric acid cycle. This pathway will be dealt with in more detail later (Sections 4.3 and 4.6).
>
> ---

So in whatever position the glucose is labelled, the carbon dioxide produced would be expected to be labelled.

However, the carbon released as CO_2 from the reaction pyruvic acid→acetyl CoA+CO_2 would be expected to be produced first, and so the *initial* rate of labelled *CO_2 production can be used to indicate the nature of the metabolic pathways. Look at Figure 2. Use marks on the carbons, e.g.* to determine in which order the carbon atoms will be released.

ITQ 1 (a) Which of the following describes the relationship between the *initial rates* of release of *CO_2 for the different positions of label:

A $1>3>4>6$
B $(1=3)>(4=6)$
C $6>4>3>1$
D $(1=6)>(4=3)$
E $(4=3)>(1=6)$
F $1=3=4=6$

The symbol > means greater than. So $1>3$ means that more label is found in CO_2 from 1-(^{14}C)-glucose than from 3-(^{14}C)-glucose.

(b) Now do the same thing for the pentose pathway; but consider glucose labelled only in positions 1, 2 or 6. Which of the following now applies:

A $1>2>6$
B $(1=6)>2$
C $2>(1=6)$
D $6>2>1$
E $(1=2)>6$

Answers and comments are on p. 38.

C-1/C-6 ratio

It is possible, then to compare the metabolism of C-1 and C-6 labelled glucose. If only glycolysis is occurring, C-1 = C-6; if only the pentose phosphate pathway, C-1 > C-6. The C-1/C-6 ratio can thus be used as a measure of the level of the pentose phosphate pathway. This has been measured for a number of tissues, and figures of 5–30 per cent for the proportion of glucose metabolized in this way obtained. These results will be dealt with in more detail later (in Section 4.7). So although the pentose phosphate pathway can accomplish the complete breakdown of glucose, the majority of glucose breakdown is by glycolysis, which may then be followed by the reactions of the citric acid cycle.

The pentose phosphate pathway also provides a source of pentoses, e.g. for DNA and RNA synthesis.

Now do SAQs 1–5 (see p. 36).

4.3 The Citric Acid Cycle as a Catabolic Route

Study Comment

This Section introduces the reactions of the citric acid cycle as a continuation to the glycolytic pathway, i.e. as a catabolic route. You should bear in mind that this is only one of its functions – some of the others will be introduced later, in Section 4.6.

The properties of control enzymes of both glycolysis and the citric acid cycle will be considered in their role as parts of a system of control, rather than as individual control points.

The reactions of the citric acid cycle, one of the central pathways, are shown in the Chart. As before, substrate level phosphorylations, in this case of GDP to GTP, are shown on overlay A, oxidation reduction reactions on overlay B. Overlay C names the metabolites of the sequence of which you should be most aware. Most of these are *branch points*, i.e. metabolites from which two or more distinct pathways diverge. These pathways are also briefly indicated on overlay C.

Remember also, from Unit 3, that the citric acid cycle occurs in the mitochondria; some of the relevant enzymes are also found in the cytoplasm, but it is only in the mitochondria that the complete cycle can occur. The overall reaction of the citric acid cycle, plus the pyruvic dehydrogenase reaction may be summed up as follows:

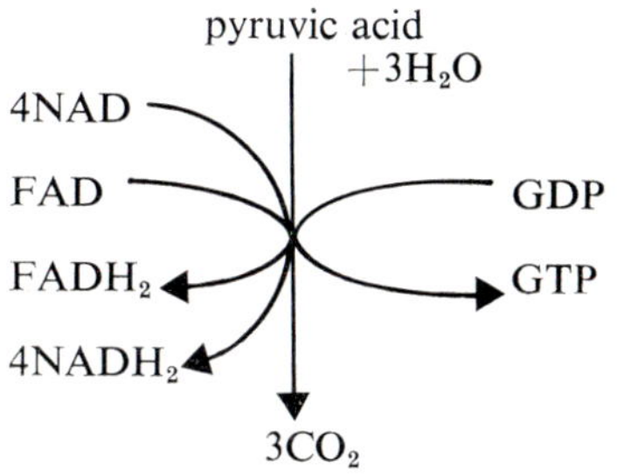

What will happen to the reduced coenzymes?

They will be oxidized by the electron transport system of the mitochondrion, eventually by molecular oxygen. This process generates ATP from ADP.

For each $NADH_2$ molecule oxidized, how many ATP molecules will be produced from ADP and phosphate (Unit 3)?

Three.

And from $FADH_2$ (which is produced by succinic dehydrogenase)?

Two.

How many ATP molecules are produced, then, for each pyruvic acid oxidized?

Fourteen, or fifteen if the GTP produced by succinic dehydrogenase is considered equivalent to ATP.

This is the equivalent of 30 ATP molecules for each glucose, because glucose→ 2 pyruvic acid.

How many ATP molecules are produced from glucose→2 pyruvic acid if the $NADH_2$ produced is odixized by conversion of pyruvic acid to lactic acid or of acetaldehyde→ethanol (i.e. under anaerobic conditions).

Two.

And if the $NADH_2$ is oxidized by the electron transport system?

Eight, two directly from glycolysis, and up to three from each $NADH_2$ oxidized (depending on how the $NADH_2$ is oxidized. See Section 4.6.3 for entry of $NADH_2$ into mitochondria).

Thus glycolysis allows the cell to use only a small fraction of the energy which can be used if the glucose molecule is fully oxidized to carbon dioxide and water.

Another way of saying this, is that an aerobic cell, using glucose, is getting most of its energy from the operation of the citric acid cycle and oxidative phosphorylation acting cooperatively.

4.3.1 The control points of pyruvic acid metabolism

Some generalizations about controlling enzymes were made in Section 4.2.2:

1 They catalyse an *early* reaction in the metabolic sequence, frequently the first specific to that pathway.

2 They are inhibited by metabolites that can be considered to be *products* of the pathway and/or activated by metabolites that are *starting metabolites* or *precursors* of the pathway.

If citric synthetase is considered as the first step of the citric acid cycle, then point 1 is satisfied, as the enzymes that are known to be important in regulating pyruvic acid metabolism within the mitochondrion are pyruvic dehydrogenase, citric synthetase and isocitric dehydrogenase. Look at the Chart with overlay C to remind yourself of the reactions catalysed by these enzymes. Pyruvic dehydrogenase and isocitric dehydrogenase have been dealt with to some extent in Unit 3. Pyruvic dehydrogenase is not, of course, strictly a part of the citric acid cycle, but it is so closely associated with it that it is relevant to consider it here.

The pyruvic dehydrogenase system and citric synthetase are both inhibited by ATP, which clearly fits point 2 above.

isocitric dehydrogenase

Isocitric dehydrogenase is more complex. Most organisms contain two types of isocitric dehydrogenase, one linked to NAD, the other linked to NADP, as mentioned in Unit 3 (Section 3.7.6). The NAD-linked enzyme is found only in mitochondria, while the NADP-linked enzyme is found both in mitochondria and cytoplasm. It was long thought that the NADP-linked enzyme was the main one, because its activity in mitochondrial preparations was much higher. This conclusion was invalidated when it was found that ADP is a specific, highly effective activator of the NAD-linked enzyme, previous studies having been carried out in solutions containing no ADP. Neither ATP nor AMP can substitute for this activation; indeed, ATP is an inhibitor. The NAD-linked enzyme is now known to be the one involved in the citric acid cycle. The enzyme is also activated by NAD and inhibited by $NADH_2$ thus completing a very neat pattern of control.

The NADP-linked enzyme is now believed to have a mainly cytoplasmic function, being involved in one of the 'shuttles' that you will come across in Section 4.6.3.

4.3.2 Control processes – the Pasteur effect

So far regulation has been considered in terms of single control points. Although these have been related to their pathways, in that precursors and products have been defined according to a broad view of metabolism and not just as the immediate precursors and products of that reaction, no attempt has been made to study the way in which the various types of control interact. This must now be remedied. But first, let us consider a very basic regulating influence on reaction rates, the effect of availability of reactants or *mass action* and ignore the effects of regulating enzymes – assume that all enzyme activities remain constant.

mass action

Consider once again the stoicheiometry of anaerobic glycolysis:

ATP consumption

The cell, of course, will be using ATP for many things – maintenance of membrane concentration gradients, the turnover of cell components, motility, net synthesis for growth, or secretion, perhaps. The limiting factors of glycolysis are the availability of ADP, glucose, and NAD. Assume the glucose level remains constant. Glycolysis generates very little ATP for each glucose molecule used, and so it is likely to be used up as fast as it can be produced. ADP will be plentiful, or to put it another way, the ATP/ADP ratio will be low, and so glycolysis will occur maximally, or nearly so. Now suppose air is allowed to reach the cell, and it therefore becomes aerobic. The stoicheiometry will now be:

ATP/ADP ratio

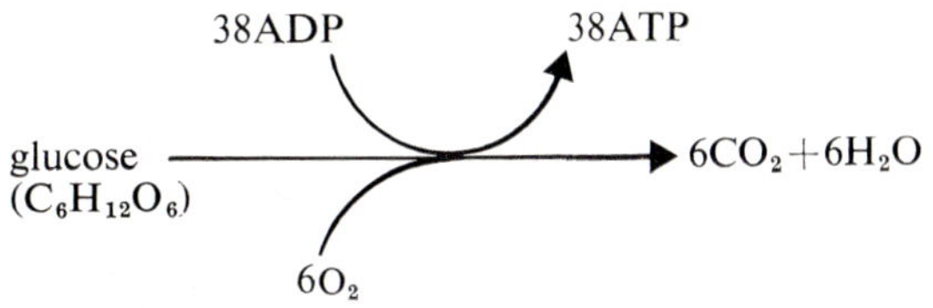

Assume that the ATP consumption remains the same. What will happen to the ATP/ADP ratio?

It will rise.

What will happen to the rate of glucose consumption?

It will fall, to the point where ATP production is once again the same as consumption, i.e. to 2/38 of its former value.

Clearly the change in ATP/ADP ratio necessary to produce this change in rate of glucose metabolism by mass action alone would be very large.

It would be more realistic to assume that the ATP-using reactions are themselves subject to mass action laws in this artificial model, in which case they would increase, and there would be a smaller, but still very substantial increase in ATP/ADP ratios, and decrease in rate of glucose metabolism following the transition from anaerobic to aerobic conditions.

Now, the decrease in rate of glucose metabolism on transfer of an organism, or tissue from anaerobic to aerobic conditions is well established. It is called the *Pasteur effect*. What does *not* occur is the very great change in ATP/ADP ratios predicted, or the overall increase in ATP-using reactions – although specific ones, e.g. glycogen synthesis, may be greatly stimulated. This is in large part due to the action on the regulating enzymes that have already been listed.

Consider the control points of glucose catabolism, and their activators and inhibitors (see Fig. 3).

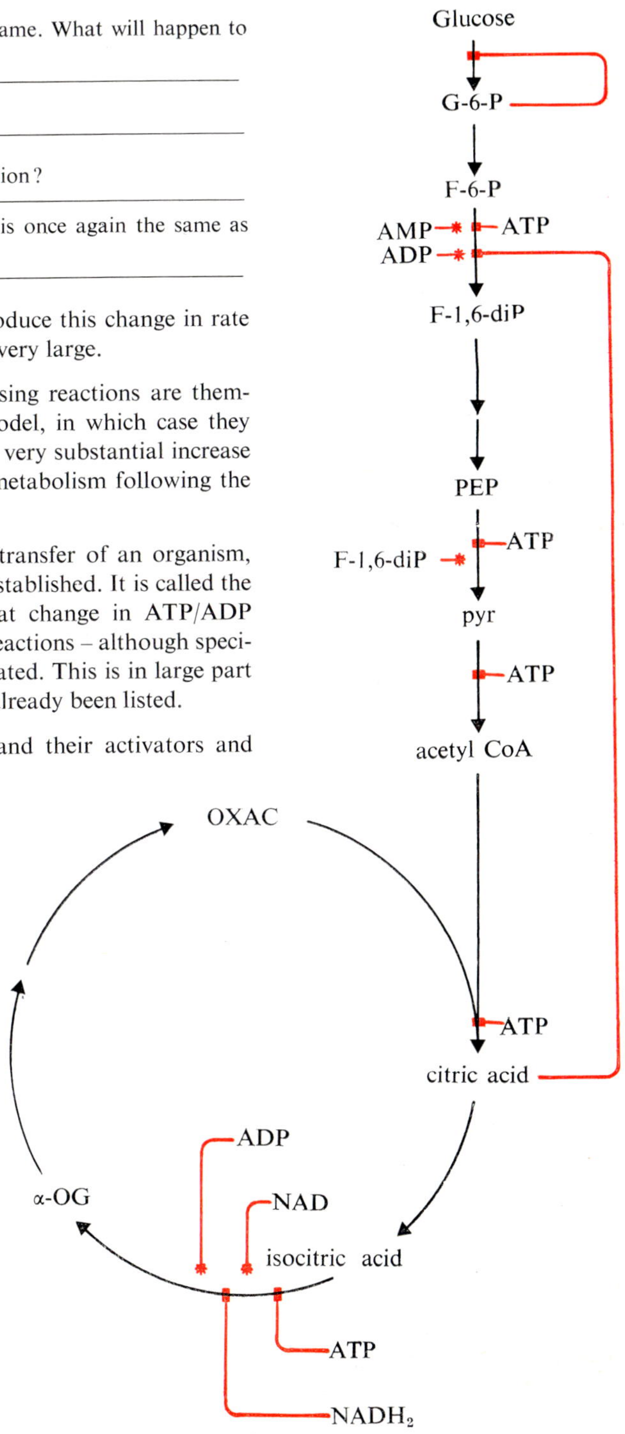

Figure 3 The control points of glucose catabolism.

KEY

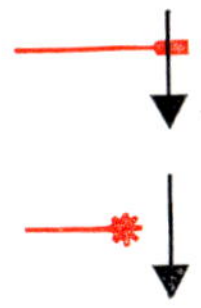

inhibition

activation

G-6-P	glucose-6-phosphate
F-6-P	fructose-6-phosphate
F-1,6-diP	fructose-1,6-diphosphate
PEP	phosphoenolpyruvate
pyr	pyruvic acid
α-OG	α-oxoglutaric acid
OXAC	oxaloacetic acid

Clearly, for each of the controlling enzymes, ATP is an inhibitor, except for hexokinase, which is inhibited by glucose-6-phosphate, its own product, which will tend to build up if phosphofructokinase is inhibited. Two of the enzymes have ADP as an activator. Thus the effect of a rise in the ATP/ADP ratio in the cell is in the same direction as that of the mass action effect, to slow down glucose metabolism. The significance of these regulatory effects is that the rate of glucose metabolism can respond to a smaller change in ATP/ADP ratios. In this way, the ATP level of the cell tends to be *buffered.* A similar argument applies to the ATP-using reactions, glycogen synthesis for example. Glycogen synthetase can be activated by glucose-6-phosphate, the starting material of glycogen synthesis. So again, a smaller change in cell composition can regulate metabolism than would be necessary if only mass action operated. This buffering of the cell composition, and in particular of ATP/ADP ratio, is important because of the dependence of the cell on ATP-using processes that must continue at all times, such as maintenance of ionic equilibria and continued turnover of components.

Two other controlling factors of the pathway have not been mentioned, the inhibition of phosphofructokinase by citric acid and the inhibition of isocitric dehydrogenase by $NADH_2$ and its activation by NAD. The major importance of these controls would seem to be under aerobic conditions, in adjusting the level of glycolysis to that of the citric acid cycle, and of the citric acid cycle to the activity of mitochondrial oxidative phosphorylation.

Now do SAQ 6 (p. 36).

4.4 Fatty Acid Breakdown

Another catabolic pathway which leads into the citric acid cycle is the breakdown of fatty acids. This Section is considered in a more historical fashion than the previous ones, because some of the early experimental work produced clear-cut results which were later verified by more sophisticated modern techniques. Fatty acid breakdown is a very important energy-producing reaction, particularly so, in mammals, in tissues like muscle, liver and kidney cortex. Under fasting conditions, fatty acid oxidation becomes the major source of energy in these tissues.

triglycerides

Fats are stored in the body of the animal mainly in adipose tissue, as triglycerides which are glycerol esters of fatty acids, as shown in Figure 4. Palmitic acid (Fig. 4c) is the commonest fatty acid in mammals. (The numbering system in Fig. 4c is given in connection with the structured exercise in Section 4.7.)

CH_2OH
|
$CHOH$
|
CH_2OH

(a) glycerol

$CH_2O—OC.(CH_2)_{14}CH_3$
|
$CH\ O—OC.(CH_2)_{14}CH_3$
|
$CH_2O—OC.(CH_2)_{14}CH_3$

(b) glycerol-tripalmitate

16 15 14 13 12 11 10 9 8 7 6 5 4 3 2 1
$CH_3.CH_2.CH_2.CH_2.CH_2.CH_2.CH_2.CH_2.CH_2.CH_2.CH_2.CH_2.CH_2.CH_2.CH_2.COOH$

(c) palmitic acid

Figure 4

lipolysis

The first stage of the breakdown is the hydrolysis of these esters to glycerol and free fatty acids, which is known as *lipolysis* and occurs mainly in adipose tissue. The fatty acids are then released into the bloodstream, and broken down mainly in liver muscle and kidney cortex.

Plants and bacteria also contain fats, oils, or related compounds, and here again, lipolysis must precede fatty acid oxidation. The glycerol produced is metabolized via pyruvic acid.

coenzyme A

There is an initial activation reaction in the sequence of fatty acid oxidation in some respects analogous to that of glycolysis. However, in this case the reaction is not a phosphorylation, but an esterification with coenzyme A. This coenzyme has a central function in metabolism as an acyl group carrier, which may be considered as being analogous to the phosphate-carrier function of ATP and creatine phosphate. Coenzyme A is represented in chemical equations as CoA.SH, as its active group (i.e. the one which combines with the acyl group) is a sulphydryl group, —SH.*

* *The reaction is thus strictly a* thio*esterification.*

The reaction of the fatty acid with coenzyme A is linked to ATP breakdown, with the following stoicheiometry:

$$RCOOH + ATP + HS.CoA \rightarrow R{-}\overset{\overset{\displaystyle O}{\|}}{C}{-}S{-}CoA + \text{pyrophosphate}$$

All the subsequent reactions occur to coenzyme A derivatives.

In 1904, Knoop did a series of experiments in which he fed fatty acids 'labelled' with phenyl groups to dogs, and observed the compounds excreted in the urine. He found the following:

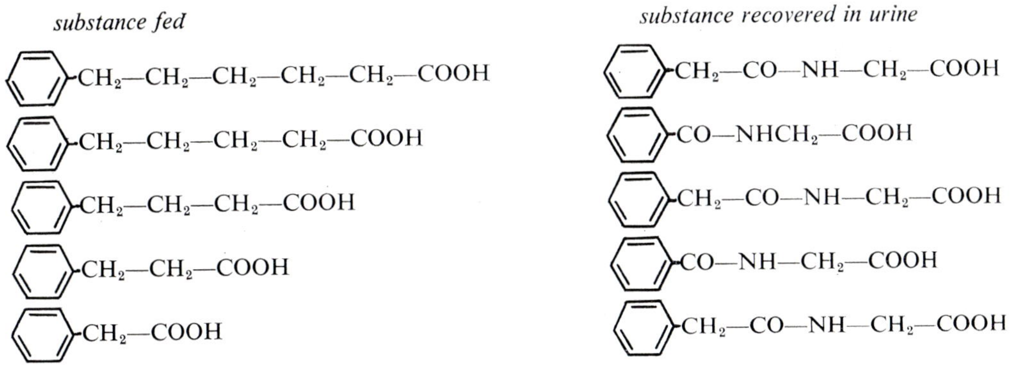

Before looking at the structured exercise below (which Knoop did not have!) count the number of carbon atoms in each compound, and try and suggest an important feature of fatty acid metabolism. (We have not included intuitive leaps in the objectives, because they are hard to assess, and harder still to teach.) Don't spend too long before going on.

ITQ 2 Which of the following conclusions are supported by the above data:

1 Adding a phenyl group has no effect on fatty acid metabolism.

2 The phenyl fatty acid is metabolized to a compound that is combined with glycine ($NH_2.CH_2.COOH$) before being excreted.

3 All fatty acids are metabolized to the same final product.

4 Even-numbered fatty acids are metabolized by a completely different route to odd-numbered fatty acids.

5 Fatty acids are broken down by successive removal of 2-carbon fragments.

6 Fatty acids are broken down by insertion of a nitrogen atom into the chain.

Answers and comments are on p. 38.

The sequence of reactions in fatty acid breakdown is as follows:

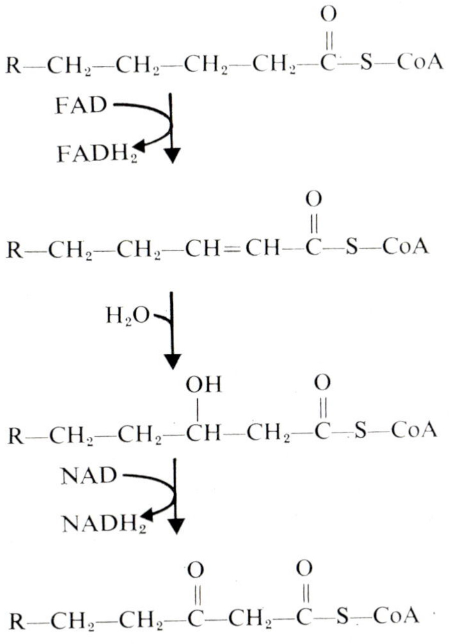

So the β-oxidation involves two oxidative steps. The next reaction involves the **β-oxidation**

cleavage of the carbon chain by a second molecule of coenzyme A, a process known as *thiolysis* (analogous to hydrolysis):

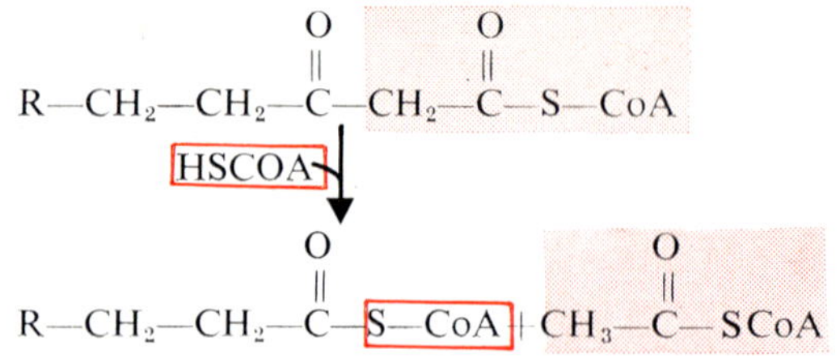

acetyl CoA

The sequence is then repeated, shortening the fatty acid by two carbon units each time.

What would you expect the final product to be in the case of a 22-carbon fatty acid?

---

11 molecules of acetyl CoA

What would the final product be for a 19-carbon fatty acid?

---

9 molecules of acetyl CoA and 1 molecule of the compound

$$CH_3\text{–}CH_2\overset{\overset{\large O}{\|}}{C}\text{–S–CoA}$$

(propionyl CoA) (You may also have suggested a one-carbon compound of some kind. This is a reasonable suggestion, but never, in fact, occurs.)

The propionyl CoA can then be metabolized to succinyl CoA. You should note that, although some odd-numbered fatty acids do occur, even-numbered ones are much commoner.

Now do SAQ 7.

4.5 Protein Breakdown

Study Comment

This Section is a rather brief account of a complex subject. A fairly detailed account is given in *Biochemistry* by A. L. Lehninger* pp. 434–40, which is black page reading.

The routes for protein catabolism begin with the complete enzymatic hydrolysis of the protein to amino acids. This may occur extracellularly, as when the protein is in food, examples being the proteolytic enzymes of the digestive tract, or those excreted by some micro-organisms; or intracellularly, in the normal turnover of proteins, or during the net protein catabolism that occurs during fasting. The individual amino acids may be synthesized into new protein molecules, or themselves catabolized. Each of the 20 amino acids is catabolized by a separate pathway, each consisting of a number of steps. These will not be listed, but they generally end up either with pyruvic acid, acetyl CoA, or with one of the citric acid cycle intermediates. Several of these routes are indicated on overlay C, in particular those for the five amino acids that have fairly direct assess to the citric acid cycle, alanine, glutamic acid, glutamine, aspartic acid and asparagine. Check the Chart with overlay C – you should know the routes of entry for these five amino acids and the corresponding keto acids.

At some stage in amino-acid catabolism, the amino group must be removed. In the case of glutamic acid, this occurs by *oxidative deamination*, the enzyme **oxidative deamination**

* *Lehninger, A. L. (1970),* Biochemistry, *Worth.*

being glutamic dehydrogenase. The reaction is:

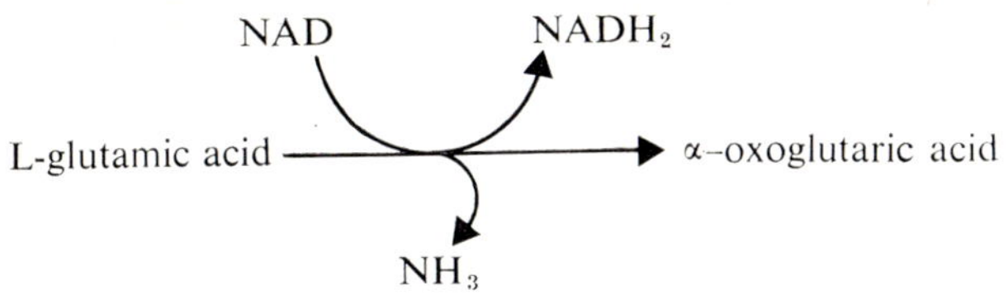

transamination

In all other cases, the amino group is lost by transamination with a keto acid, frequently α-oxoglutaric acid, the enzyme being glutamic transaminase, although there are a number of other transaminases. A typical reaction catalysed by glutamic transaminase is:

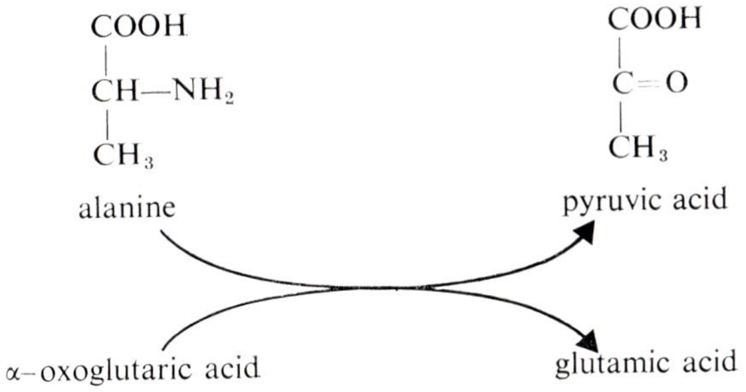

In this way glutamic acid becomes a 'channel' for amino groups, being the only amino acid that can be deaminated at any great rate.

You will not be surprised to learn, then, that glutamic dehydrogenase is a regulatory enzyme. It is inhibited by ATP and $NADH_2$, and activated by ADP. In this way, deamination, and therefore catabolism of amino acids, is stimulated by low ATP/ADP ratios.

Under conditions where substantial amino-acid catabolism is occurring, ammonia will be produced continuously. However, it is toxic, and must be got rid of.

Different animals have different mechanisms by which the ammonia is removed. In aquatic animals free ammonia is released in solution into the surrounding water. Most terrestrial animals have a detoxication mechanism that allows the waste product to accumulate to some extent. In mammals, the ammonia is normally converted to urea

urea

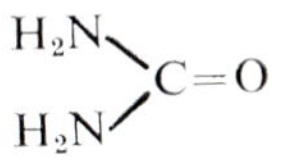

in the liver before being excreted.

4.6 The Citric Acid Cycle in Anabolism. Some Associated Central Pathways

Study Comment

In this Section the role of the citric acid cycle as a source of material for anabolism is considered, and some associated reactions that are necessary for it to function in this way. As these reactions vary considerably among different types of organism, they have not been included in the general scheme but will be dealt with separately. The compartmentation of the citric acid cycle will also be dealt with here, as the associated reactions are also concerned with entry and exit of metabolites to and from the mitochondria.

The citric acid cycle has a role in the catabolism of all the major cell components. However, its intermediates are also the source of materials for synthesis, as indicated on overlay C. Thus of the components in and around the citric acid cycle, oxaloacetic acid, α-oxoglutaric acid, succinic acid, acetyl CoA, and pyruvic acid form starting points for major synthetic pathways.

In the citric acid cycle, as usually written (abbreviated form),

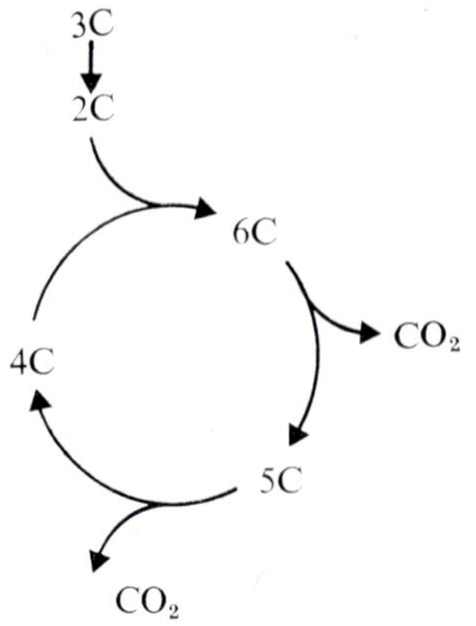

the role of the cycle intermediates is catalytic—they cannot change in level. Clearly this cannot be the case if the cycle is to act either as the source of materials for synthesis, or as a means for the net breakdown of substances whose pathways end with one of the cycle intermediates. There must be some means of generating and breaking down the cycle intermediates that is independent of these routes. In animals, this is carried out by the carboxylation of pyruvic acid to oxaloacetic acid:

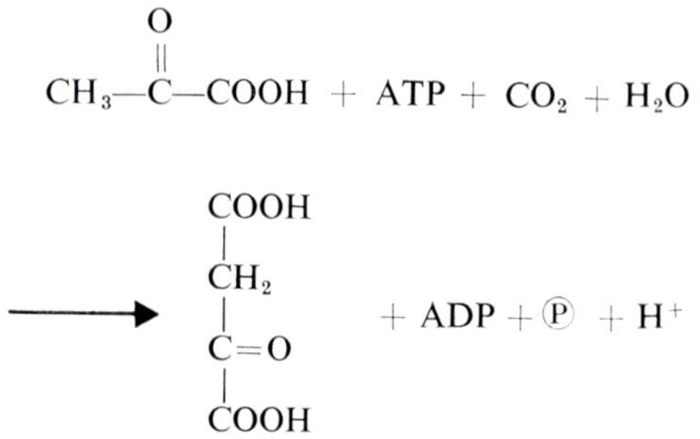

pyruvic carboxylase

This reaction is freely reversible, and can act either to generate oxaloacetic acid from pyruvic acid to maintain the level of the cycle intermediates, or convert oxaloacetic acid to pyruvic acid, by which means all the components of the cycle can be broken down. This kind of synthetic pathway has been called an anaplerotic ('replenishing') pathway.

anaplerotic pathways

Note that pyruvic carboxylase is also the first reaction of gluconeogenesis from pyruvic acid, and it is followed by a decarboxylation:

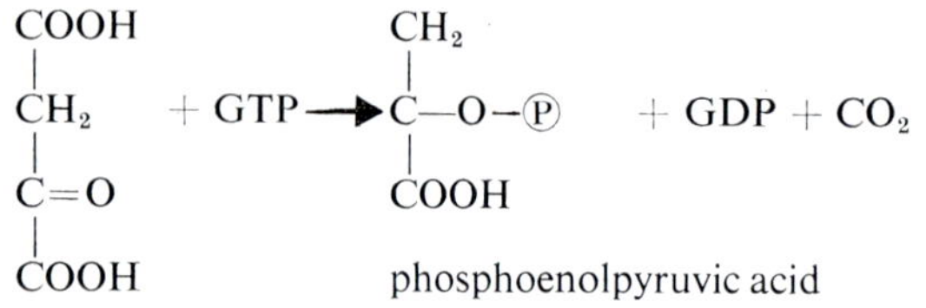

Gluconeogenesis, however, occurs in the cytoplasm, while the citric acid cycle is mitochondrial.

Interconversion of pyruvic acid and oxaloacetic acid occurs in the mitochondrion (see Section 4.6.3).

4.6.1 The glyoxylic acid sequence

In higher plants and many micro-organisms, there is an alternative sequence known as the glyoxylic acid sequence. The first reaction involves the splitting of isocitric acid to produce succinic acid and glyoxylic acid:

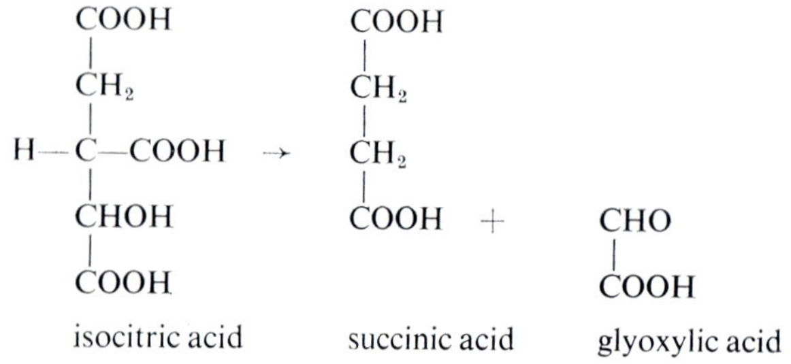

The glyoxylic acid then reacts with a molecule of acetyl CoA to give malic acid:

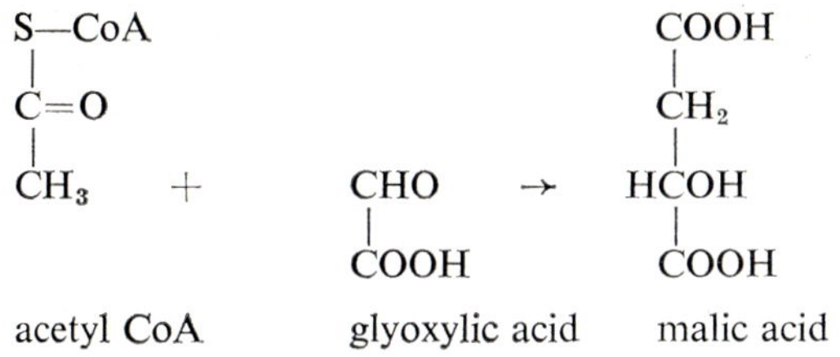

Thus one molecule of a citric acid cycle intermediate and one of acetyl CoA gives two molecules of citric acid cycle intermediates. Refer to the Chart with overlay C.

You may have noticed that, given the pathways described so far, for animals, there is no route by which *net synthesis* of pyruvic acid can occur from acetyl CoA. (However, it is possible for radioactive label from acetyl CoA to get into pyruvic acid, via oxaloacetic acid. You should recognize that exchange of a radioactive atom does not *necessarily* involve a net synthesis.) This implies that only pyruvic acid, and not acetyl CoA, can be used as a starting material for gluconeogenesis, and that the citric acid cycle intermediates can be generated only from pyruvic acid, and not from acetyl CoA. This is indeed the case. However, the situation is quite different for those plants and micro-organisms capable of the glyoxylic acid sequence. In this case, not only can the citric acid cycle intermediates be generated from acetyl CoA, but the conversion of oxaloacetic acid to phosphoenol pyruvic acid allows acetyl CoA to act as the starting material for glucose synthesis.

4.6.2 Ketosis

The inability of animals to maintain the citric acid cycle intermediates from acetyl CoA has an important consequence. What?

The cell cannot utilize acetyl CoA, or fatty acids, as sole source of energy. So some carbohydrates or proteins must be catabolized to maintain cell functions.

What will happen if this is not the case?

There will be a build-up of acetyl CoA, or of its derivatives.

In fact three compounds, acetoacetic acid, β-hydroxybutyric acid and acetone accumulate. These three compounds, known as *ketone bodies*, are derived from acetyl CoA in the following way: **ketone bodies**

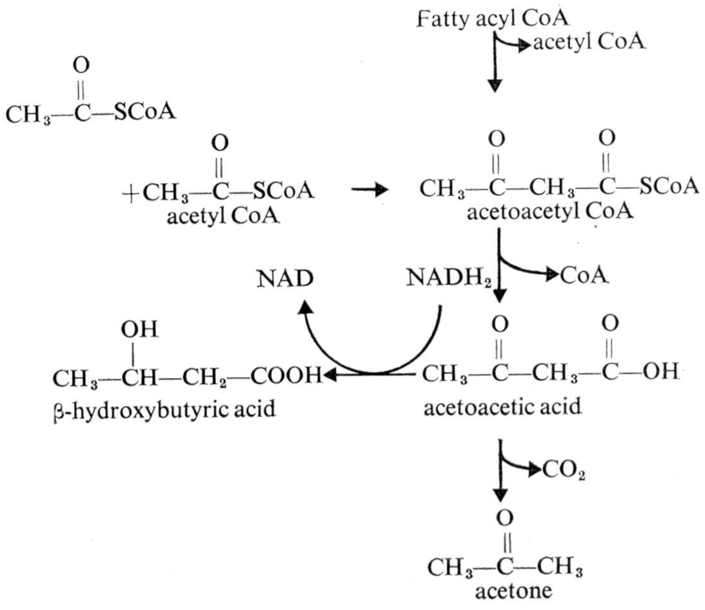

These compounds are normal metabolic products, and can be utilized by most tissues in the mammalian body. However, under conditions where the amount of carbohydrate catabolized is insufficient to balance the amount of fatty acid being catabolized, these ketone bodies will accumulate, a condition known as *ketosis*.

This constraint is important from a nutritional point of view, both because under these conditions catabolic efficiency is impaired, and because the substances that build up are toxic. In particular, the acid/base balance of the body is disturbed.

Mild ketosis is common in people or animals on a high fat diet, or (what is in effect similar) in obese people on extended fast to lose weight. Extreme ketosis can occur in untreated diabetes. This disease consists of a deficiency of the hormone insulin (see Section 4.8.1) which is necessary for efficient uptake of glucose by cells. Under these conditions, there is simultaneously an abnormally high reliance on fat metabolism and, under most conditions, a very high level of blood glucose. Treatment is in principle extremely simple – the regular administration of insulin by injection. In untreated diabetes ketosis results, and the disturbance in acid-base balance can be sufficient to cause death.

diabetes

insulin

The ability of proteins to replace carbohydrate in allowing the efficient metabolism of acetyl CoA is due to the fact that many of the amino-acid breakdown pathways end either in pyruvic acid or in citric acid cycle intermediates. Such amino acids can also act as precursors for gluconeogenesis, and have therefore been classified as *glucogenic*. Other amino-acid breakdown pathways end in acetyl CoA – these are termed *ketogenic*. The presence of the glyoxylic acid sequence in plants is significant in that it allows, for example, seeds to rely entirely on fat or oil as an energy source, a given mass of fat storing more energy than the same mass of carbohydrates or protein.

4.6.3 Mitochondrial permeability

The localization of a number of important reaction sequences in the mitochondrion, while others occur in the cytoplasm, has been referred to several times, the localization of each pathway having been mentioned in the corresponding Section. The question of mitochondrial permeability must now be considered, because a number of important compounds do not penetrate the mitochondrial membrane, for instance $NADH_2$, which is produced by glycolysis in the cytoplasm (as well as by the citric acid cycle in the mitochondrion). How can the $NADH_2$ produced in the cytoplasm be oxydized by the mitochondrial electron transport system? Fatty acid breakdown is mitochondrial, but fatty acyl CoA cannot penetrate the mitochondrial membrane. Similarly the principal route of fatty acid synthesis, as will be seen, occurs in the cytoplasm, but the source of acetyl CoA is mitochondrial. Another example is the initial reaction of gluconeogenesis from pyruvic acid, the pyruvic carboxylase reaction producing oxaloacetic acid which again cannot cross the mitochondrial membrane to the cytoplasmic enzyme system* responsible for the rest of the sequence. These reactions are believed to be carried out by reaction sequences which involve the conversion of the impermeable compound to a permeable one, followed by reconversion after penetration. Examples of suggested mechanisms for the reactions outlined above are given in Figures 5, 6 and 7.

Now do SAQ 8.

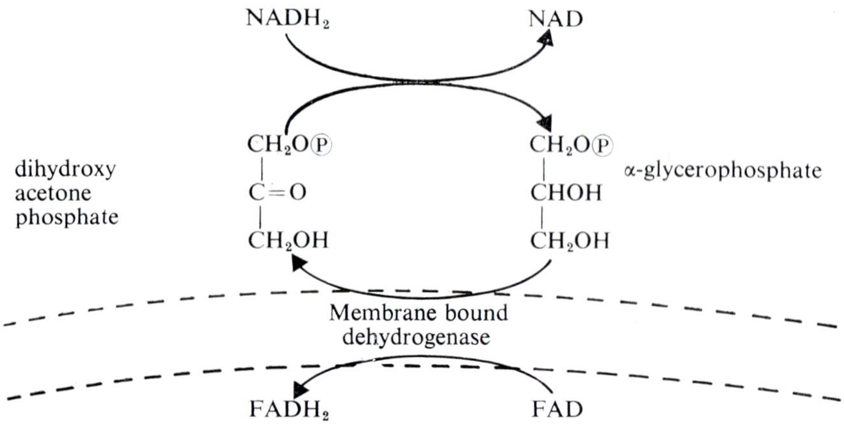

Figure 5 In the case of NAD, a pair of compounds which can react with a membrane-bound enzyme system, α-glycerophosphate (α-GP) and dihydroacetone phosphate (DHAP) are utilized. The $NADH_2$ reduces the DHAP to α-GP. The α-GP then reacts with the membrane-bound enzyme where it can reduce mitochondrial FAD to $FADH_2$ by a reverse of the previous reaction. (You should note that the oxidation of $NADH_2$ is linked to the reduction of FAD. So the oxidation of each molecule of $FADH_2$ results in two *not three molecules of ATP being produced from ADP.)*

* *Phosphoenol pyruvate carboxykinase, the enzyme responsible for the conversion of oxaloacetate to phosphoenol pyruvate, varies in its subcellular distribution. So the details of the system described will vary from animal to animal.*

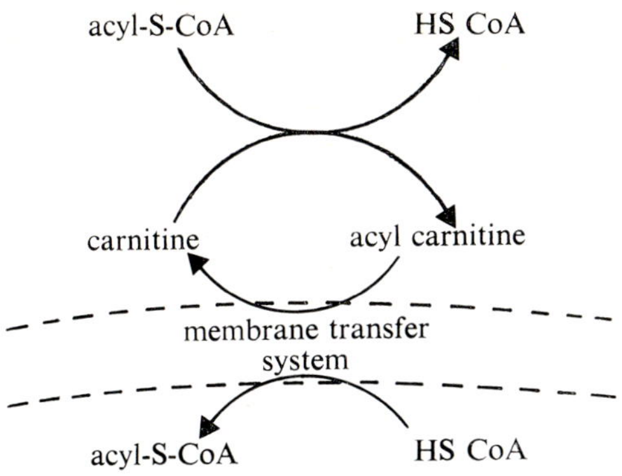

Figure 6 (above) In the case of acyl CoA, a compound called carnitine substitutes for CoA reversibly, the acyl carnitine being able to penetrate the membrane. It is believed that the carnitine has only this function of acting as a carrier for acyl groups. It is equally effective for acetyl CoA and for fatty acyl CoA compounds.

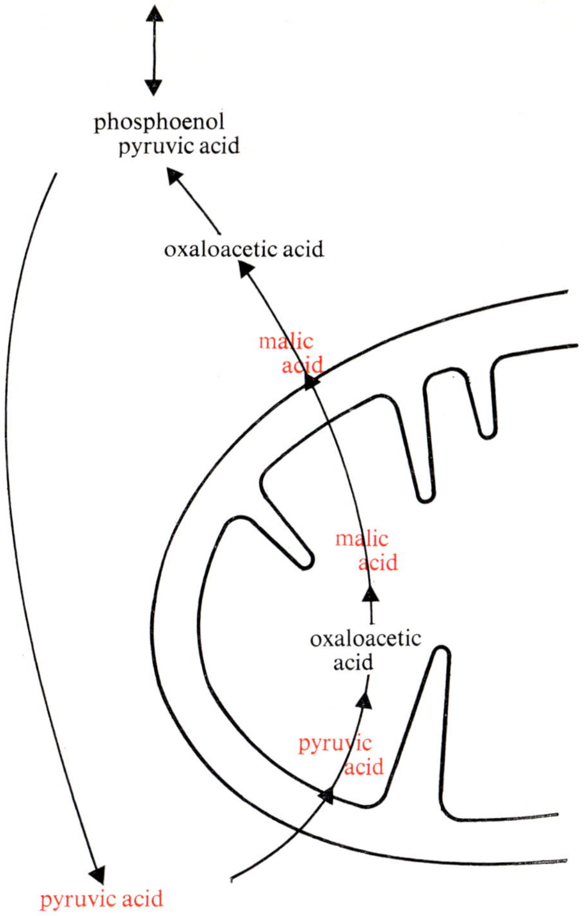

Figure 7 (right) In this reaction, the oxaloacetic acid formed within the mitochondrion is converted to malic acid, which can penetrate the mitochondrial membrane before being reconverted to oxaloacetic acid.

4.7 Fatty Acid Synthesis

Study Comment

In this Section a structured exercise, based on experimental data from a number of sources, is used to introduce the sequence for fatty acid synthesis.

Like glucose or glycogen synthesis from pyruvic acid, the synthesis of fatty acids from acetyl CoA does not occur by a reverse of the breakdown process. In this case, the reaction occurs by a completely separate route, although it is related to the breakdown 1oute in that the chain is built up in 2-carbon units.

The following scheme has been proposed for fatty acid synthesis, based on the studies of Lynen, Wakil and others.

$$\underset{\text{acetyl CoA}}{CH_3-\overset{\overset{\displaystyle O}{\|}}{C}-S\,CoA} + CO_2 + ATP$$

$$\longrightarrow \underset{\text{malonyl CoA}}{\overset{COOH}{\overset{|}{CH_2}}-\overset{\overset{\displaystyle O}{\|}}{C}-S\,CoA} + ADP + Ⓟ^-$$

(an initial 'activition reaction')

acetyl CoA carboxylase

Both acetyl CoA and malonyl CoA then react with another sylphydryl compound, this one a protein, called acyl carrier protein (ACP).

$$CH_3-\overset{\overset{\displaystyle O}{\|}}{C}-S\,CoA + HS-ACP$$

$$\longrightarrow CH_3-\overset{\overset{\displaystyle O}{\|}}{C}-SACP + CoASH$$

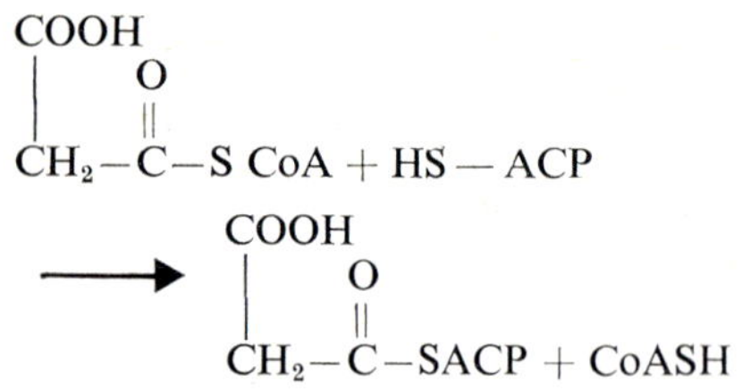

Then the two ACP derivatives react:

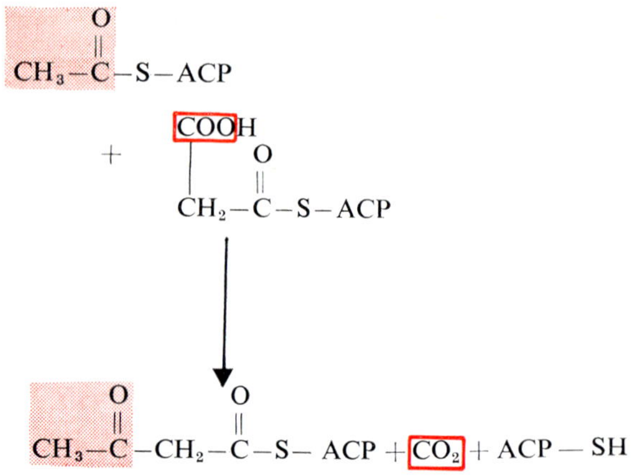

Then occur two reductions, almost identical to the reverse of β-oxidation, but in this case linked to *NADP* not to NAD or flavin, giving

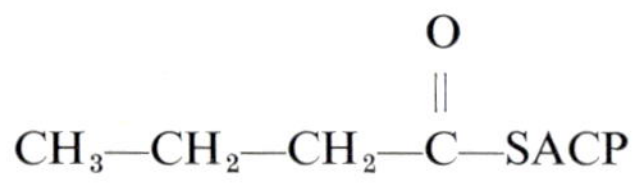

This 4-carbon acyl-ACP compound can then react with a further *malonyl* CoA and the whole sequence repeats until the full length of the chain is built up.

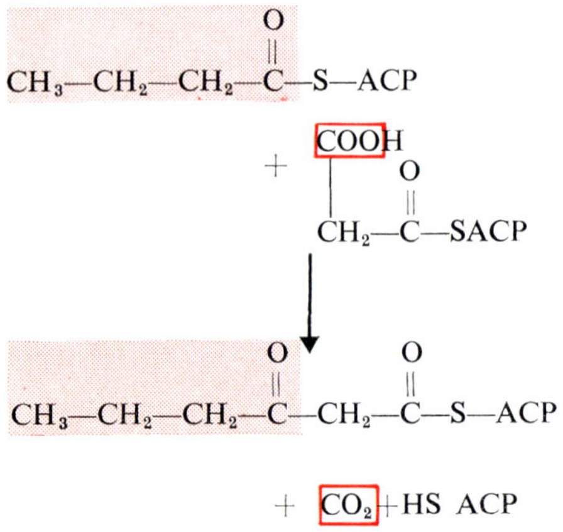

Look at this set of reactions carefully; a set of experiments to test this scheme is given. They are based on the actual experimental work which led to this suggested scheme. For each experiment, several possible results are listed.

Experiment
In a cell free system, incorporation of labelled acetyl CoA and malonyl CoA was attempted.

ITQ 3 Assign *each* possible result as one of the following:
1 Supporting the scheme above.
O Refuting, or weighing against the scheme.
X Neither supporting nor weighing against the scheme.

Results
(a) No incorporation of any precursor was achieved.
(b) Acetyl CoA was incorporated into fatty acid, but malonyl CoA was not.
(c) Malonyl CoA alone was incorporated into fatty acid.
(d) Malonyl CoA was incorporated into fatty acid only in the presence of acetyl CoA.

Experiment
Labelled acetyl CoA and excess unlabelled malonyl CoA were incorporated into palmitic acid, a straight chain, 16-carbon acid by the above system, which was then assayed for label in different positions. (For the numbering system for palmitic acid, see Fig. 4c on p. 18.)

ITQ 4 Assign each of the results below in the same way as for *ITQ 3*.

Results
(a) No label was found in the palmitic acid.
(b) Label was found uniformly distributed in the palmitic acid.
(c) Label was found only in the 1 and 2 positions.
(d) Label was found only in the 15 and 16 positions.

Experiment
Acetyl CoA was incorporated into fatty acid in the absence of malonyl CoA, but in the presence of $^{14}CO_2$.

ITQ 5 Assign each of the results below in the same way as for *ITQs 3* and *4*.

Results
(a) ^{14}C was found in all the odd-numbered carbon atoms except 15.
(b) ^{14}C was found in all the even-numbered carbons except 16.
(c) ^{14}C was not incorporated into palmitic acid.

Answers and comments to *ITQ*s *3–5* are on p. 38.

4.7.1 Control of fatty acid synthesis

In the control of this pathway it is again an early enzyme in the pathway, in this case acetyl CoA carboxylase, that is the regulating step. Acetyl CoA carboxylase may be inhibited by fatty acyl CoA compounds, although data on this point are still uncertain. It is activated by citric acid and isocitric acid. In this way, fatty acid synthesis is enhanced when the level of citric acid cycle intermediates is high.

4.7.2 The role of $NADPH_2$

You will have noticed that the oxidation-reduction reactions of fatty acid synthesis, unlike its breakdown, is linked to NADP, not NAD, each 2-carbon unit that is added to the chain requiring 2 molecules of $NADPH_2$. $NADH_2$ cannot substitute for this requirement, indeed, as stated in Unit 3, nearly all reactions that are linked to the oxidation or reduction of one of these cofactors are quite specific for one or the other.

What is the source for the $NADPH_2$ required for fatty acid synthesis? Four reactions or pathways have been covered in Units 3 and 4, which could supply this. Name them.

The pentose phosphate pathway.
Isocitric dehydrogenase.
Transhydrogenase
($NADH_2 + NADP \rightleftharpoons NADPH_2 + NAD$).
Photosynthesis.

A major source of $NADPH_2$ for fatty acid synthesis is, in animals, the pentose phosphate pathway, and in plants, photosynthesis. What might you expect about the rate of glucose oxidation by the pentose phosphate pathway in different animal tissues?

The higher the capacity for fatty acid synthesis in the tissue the higher will be the proportion of glucose metabolized by the pentose phosphate pathway.

This is what is found. Another major source of $NADPH_2$ in animal tissues, that has not been dealt with so far is the NADP-linked malic-enzyme, which catalyses the reaction:

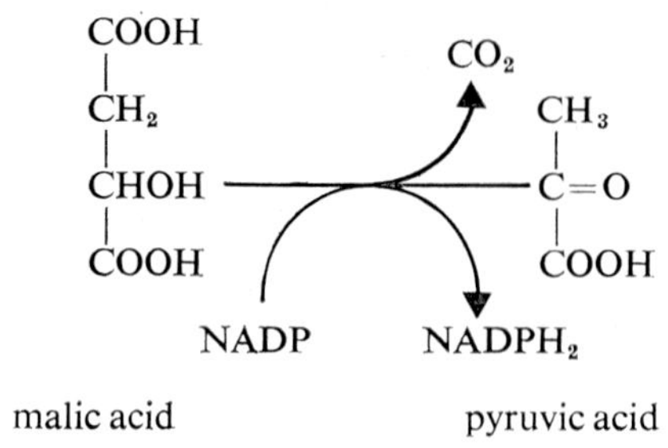

4.7.3 Other ATP-using reactions

In general, then, the catabolic reactions of the cell produce ATP, and reducing power (mainly as $NADH_2$ and $NADPH_2$), while the anabolic reactions use ATP and $NADH_2$ or $NADPH_2$. There are, of course, many other catabolic and anabolic reactions in the cell—some of these are indicated in outline on the Chart with overlay C, such as those for pyrimidines and amino acids. The synthetic pathways are again ATP-users. There are other reactions in cells which use ATP, such as muscle contraction, and maintenance of ionic balance by the cell membrane. One important synthetic pathway has not been included here, that for proteins. This will be considered further in Unit 6 of this Course.

Now do SAQ 9.

4.8 Localization and Control

Study Comment

In this Section, the localization of the pathways dealt with among the different tissues of the mammal and some general features of control are considered.

The striking thing about most of the pathways described in this Unit is their universality among living systems. Some variations have been noted, but the overall pattern is fairly universal. This may be in part due to the considerations mentioned earlier, in Section 4.2.4. And the importance of the similarities among living systems as compared to the differences is a rich source of dispute. Plants and bacteria have all sorts of pathways in addition to those found in animals, particularly synthetic pathways. For example, those amino acids, certain fatty acids, and coenzyme precursors (vitamins) that are essential to the diet of man or other animals are synthesized normally by plants, or bacteria. This means that vitamins (see Unit 2, Section 2.4.1) and other essential dietary components are an arbitrary class of chemicals and, in fact, vary from one species to another. For example, vitamin C (ascorbic acid) is only known to be required by primates and guinea-pig. Other species (or their intestinal flora) can synthesize ascorbic acid and, to them, therefore, ascorbic acid is not a vitamin. Some micro-organisms have nutritional requirements analogous to animal vitamins. In general, however, (i) they are not so clear-cut, or universal; (ii) their requirements are fewer.

The general universality of the major pathways described between species does not apply to the various tissues within the animal. Thus gluconeogenesis occurs in mammals only in liver and kidney cortex, 90 per cent of the capacity being in the liver. Fatty acid breakdown occurs mainly in liver, kidney cortex and muscle, while lipolysis occurs mainly in adipose tissue.

Fatty acid synthesis occurs mainly in adipose tissue, although some also occurs in liver. Glycolysis and the citric acid cycle, however, occur in virtually all tissues of the mammal.

The regulation of individual pathways by their metabolites has been dealt with in Sections 4.2.2 and 4.3.2. How does this form of regulation fit in with the needs of the whole animal? Different tissues must react in quite different ways to a particular change in the animal's environment.

In fasting, for instance, what would you expect to be the major reaction in adipose tissue?

Lipolysis.

In liver?

Gluconeogenesis.

Control by metabolites is unable to direct these overall functions, which are performed by a group of 'chemical messengers'.

What are these called?

Hormones. Examples are insulin, cortisone, adrenalin, thyroxin, glucagon.

4.8.1 Hormonal control

This form of control is much more versatile than metabolic control. A single messenger can produce quite different changes in different tissues, and can be produced as a response to the perception and analysis of the environment by the central nervous system. Adrenalin, for example, activates lipolysis in adipose tissue, and *glycogenolysis* (glycogen breakdown) in muscle and liver, raising the levels of free fatty acids and glucose in the blood. **glycogenolysis**

This action is opposed by insulin, which also stimulates glucose uptake by many tissues. The production of these hormones is itself controlled by the metabolic state of the animal, forming feedback loops of the sort described for thyroxin in S100[6]. The description of hormonal action in this Unit will be mainly at the level of the effects on individual tissues, and the mechanism of these effects. The way in which these hormones interact, and their significance for the animal as a whole (the physiological approach) is dealt with in Course S22–.*

4.8.2 Control of adipose tissue metabolism

Adipose tissue metabolism is frequently studied by incubation of excised epididymal fat pads** in a medium containing (^{14}C)-labelled precursor. Such a system has the advantage for studies of regulation that the system is relatively intact. In particular the membrane system can control uptake of metabolites, and so both control of uptake and intracellular regulation mechanisms can be studied.

ITQs 6 and 7
In this structured exercise, two sets of experimental data are given and a set of possible interpretations from the data. In each case you are asked to assess each interpretation as:

(1) Likely, or correct within experimental error.
(X) Not warranted.
(O) Unlikely or incorrect.

ITQ 6 *Experimental data I*
Fat pads were removed from 6 rats and incubated in the presence and absence of insulin in a medium containing 6(^{14}C)-glucose for 36h.
The following incorporations were obtained (in arbitrary units):

* *The Open University (1972) S22–* Science: A Second Level Course, Comparative Physiology, *The Open University Press.*

** *The epididymal fat pads are attached to the testis of the male mammal.*

	Without insulin	*With insulin*
Total glucose carbon uptake into tissue	280±11	840±23
glucose carbon into CO_2	69.9±5.1	207±8
glucose carbon into fatty acid	33±5.3	219±14
glucose carbon into tissue protein	6.7±0.5	11.5±0.7

Possible interpretations

1 The primary action of insulin is on transport of glucose across the cell membrane.

2 The primary action of insulin is on the enzymes of glycolysis, fatty acid synthesis, etc.

3 Insulin increases the uptake and metabolism of glucose by fat cells, but no conclusions can be drawn concerning the mechanism or site of this action.

ITQ 7 *Experimental data II*
It has been suggested (see Section 4.7.1) that acetyl CoA carboxylase is an important regulating step in fatty acid synthesis. In this experiment, also on epidydimal fat pads, 10 m*M* (^{14}C) glucose incorporations were carried out in the presence and absence of labelled or unlabelled acetate (15 m*M*). In these experiments, as well as measuring the incorporation of ^{14}C into various fractions, the contribution of the pentose phosphate pathway was measured in both cases, by a method similar to that given in Section 4.2.6. Insulin was present in these studies.

	without acetate	*with acetate*
1 total uptake of ^{14}C from glucose into tissue	20.85±2.2	24.4±1.8
2 incorporation of ^{14}C from glucose into fatty acid	8.4±0.7	9.75±0.5
3 total incorporation of ^{14}C from glucose into CO_2 (glycolysis plus pentose phosphate pathway)	7.55±0.3	10.0±0.8
4 incorporation of ^{14}C from glucose into CO_2 via the pentose phosphate pathway	4.2±0.2	7.7±0.3
5 incorporation of ^{14}C from acetate into CO_2 plus fatty acid	0±0.3	4.2±0.4

Again, a series of possible interpretations is given, and you are asked to assign them as 0, 1 or X, as above.

Possible interpretations I

1 In the presence of acetate, the total rate of fatty acid synthesis increases.

2 In the presence of acetate, the incorporation of glucose carbon into the tissue is inhibited.

3 Both acetate and glucose carbon can be incorporated into fatty acid by the tissue.

4 All of the glucose carbon taken up by the tissue is metabolized to either carbon dioxide or fatty acid.

5 In the presence of acetate, the rate of glycolysis remains unchanged.

6 More acetate carbon is taken up by the tissue in the presence of glucose than in its absence.

Now check your answers *(p. 38)* *before going on to the second part.*

Possible interpretations 2

7 All the extra glucose incorporated into the tissue in the presence of acetate is metabolized by the pentose phosphate pathway.

8 The rate of production of acetyl CoA from glucose is a rate-limiting step under these conditions.

9 The rate of oxidation of $NADPH_2$ by fatty acid synthesis is a rate-limiting step for the pentose phosphate pathway under these conditions.

10 Acetyl CoA carboxylase is a rate-limiting step under these conditions.

11 Only one of these steps will be an important control point in this system.

11 There is no single control point in this system.

4.8.3 Glycogen phosphorylase and glycogen synthetase

These two enzymes have been left until now, both because their role can best be understood in terms of whole tissue function, and because they are subject to hormonal control. The reactions they catalyse—steps in the breakdown and synthesis of glycogen, respectively—can be seen on the Chart with overlay C.

Glycogen is stored in a number of tissues, as a reserve nutrient, especially in muscle and liver. The control of its catabolism and anabolism is complex. Glycogen phosphorylase catalyses its breakdown according to the equation:

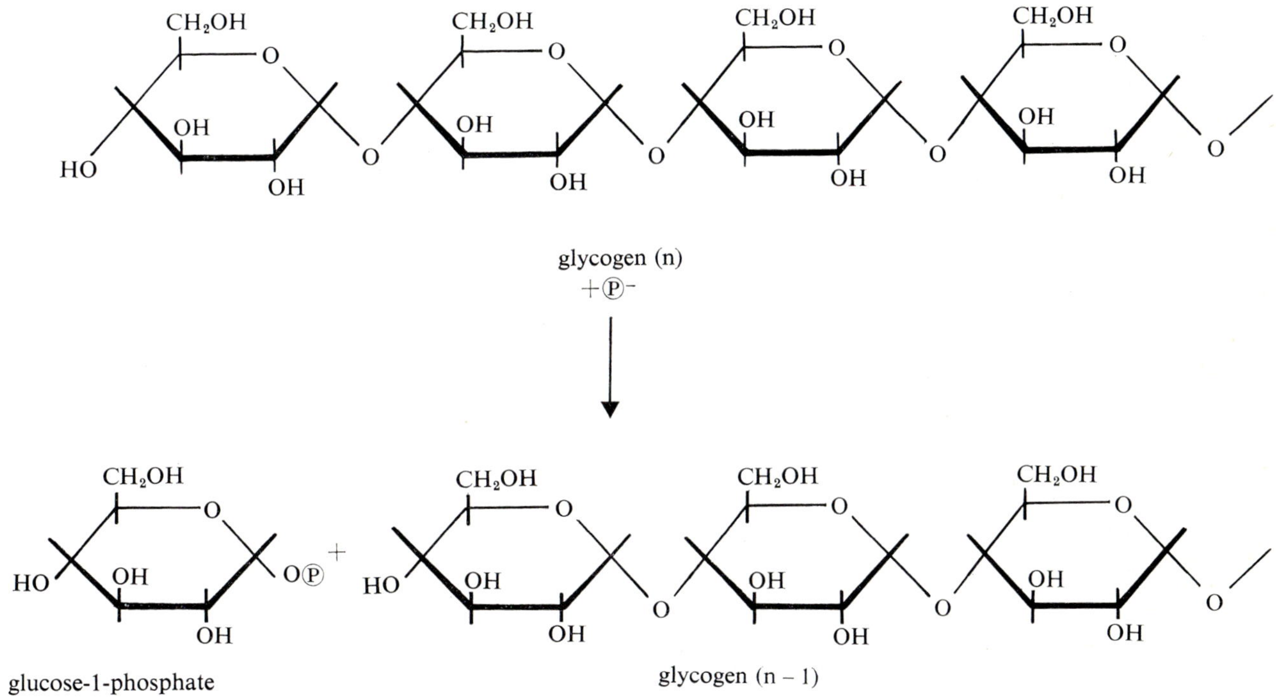

phosphorylase a and b

The enzyme that catalyses this reaction can exist in two forms: an active form, phosphorylase a, and a much less active form, phosphorylase b; these two forms can be interconverted in the cell, under the influence of hormones; adrenalin and glucagon (a polypeptide hormone from pancreas) can both stimulate the formation of phosphorylase a from phosphorylase b (activation of the enzyme) while insulin has the opposite effect. In this way, glucagon and adrenalin stimulate glycogenolysis while insulin opposes this stimulation. The enzyme, however, is also subject to metabolic control. Both forms are stimulated by AMP, but the b form requires a much higher level of AMP for maximal activity than does the a form. The normal levels of AMP are sufficient to maximally stimulate phosphorylase a, while phosphorylase b is almost inactive at these levels. It seems, however, that it is possible under physiological conditions for the level of AMP to vary sufficiently to have a controlling effect, certainly on phosphorylase b, and possibly on phosphorylase a. The significance of this metabolic control should be clear to you—high levels of AMP indicate a depletion of energy sources within the cell, and so a need for breakdown of the glycogen reserve. So the enzyme is subject to dual control, by metabolites and by hormones, and the mechanism of the two actions is quite different. Let's consider the hormonal control in more detail. The two forms of phosphorylase differ in two respects. The a form is a tetrameric molecule (i.e. composed of four subunits) which is phosphorylated on each monomer. The b form is a dimer (i.e. composed of two subunits) which lacks these phosphate groups. The interconversions themselves are enzyme-catalysed reactions; a→b is catalysed by a phosphate-removing enzyme, a *phosphatase*. b→a is catalysed by an enzyme **phosphorylase phosphatase** that transfers phosphate from ATP to the enzyme, a *kinase* which we shall refer to (for reasons that will become apparent) as kinase I. **kinase I** The reactions so far can be summed up as follows:

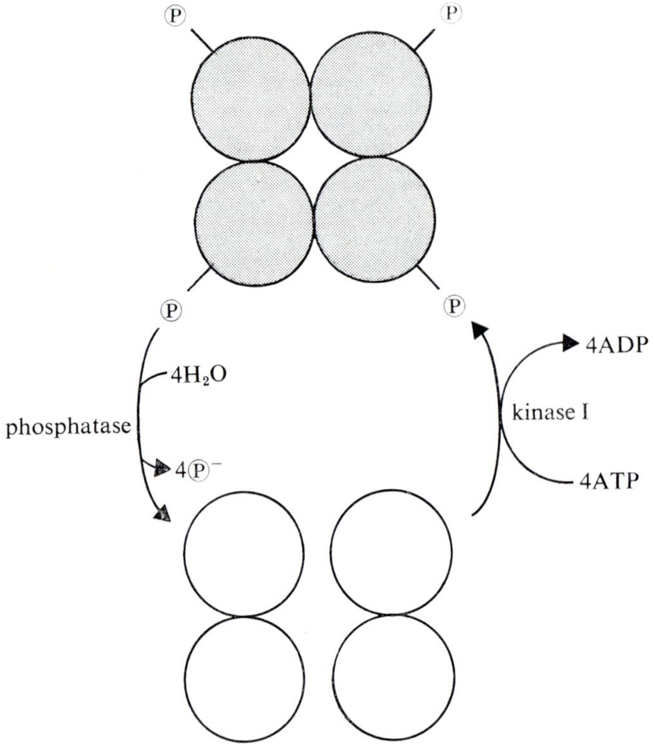

It is the kinase activity, not the phosphorylase activity, that is affected by the hormones. However, the mechanism of this activation brings a remarkable symmetry to the story, because kinase I itself exists in an activated form, which is phosphorylated, and in an inactive form, which is non-phosphorylated. The activation occurs by another kinase reaction, catalysed by a distinct enzyme which has been called 'kinase kinase', but which we prefer to call *kinase II.* **kinase II**

At least one more step, however, remains between the initial hormone reaction and the activation of phosphorylase. The hormone acts not inside the cell, but upon the cell membrane, where it activates an enzyme adenylcyclase. **adenylcyclase** This enzyme catalyses the reaction

ATP→*cyclic* AMP + pyrophosphate

The cyclic AMP is a nucleotide in which two hydroxyl groups of the phosphate group are each esterified to the ribose of adenosine (in the 3′ and 5′ positions): **cyclic AMP**

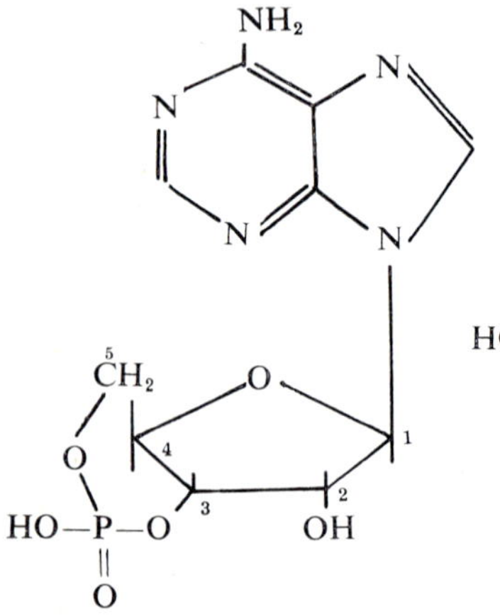

3′5′ cyclic AMP

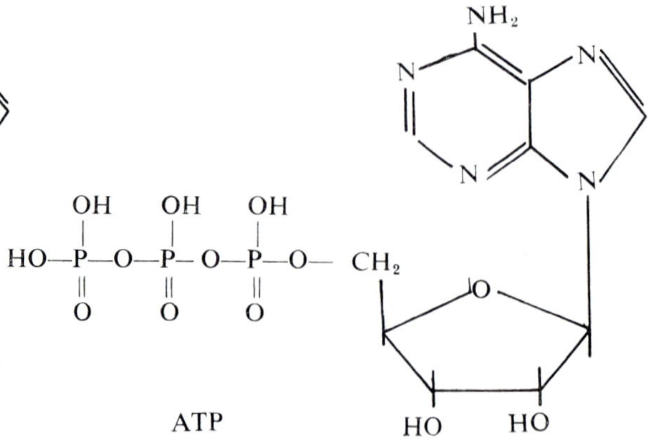

ATP

Notes

1 The prime (′) is used to distinguish the numbering of the ribose ring from that of the adenine.

2 The phosphates are shown un-ionized. At normal pH about half the hydroxyl groups will be ionized:

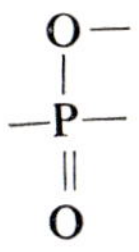

3 Do not attempt to learn these formulae. You should, however, be aware of the manner of the attachment of the phosphate group to the ribose, which is a cyclic *diester* linkage.

second messenger

Cyclic AMP has the role of a *second messenger* within the cell. Both the hormones glucagon and adrenalin stimulate its production by the membrane-bound adenyl cyclase. It is the cyclic AMP that activates the kinase II molecule, presumably by an allosteric mechanism.

enzyme cascade

This example of control has been given in detail for two reasons. One is to remind you of the complexity of the systems you are studying. The second is because it is a good example of a type of event known as a cascade. The whole activation process can be summed up in the following way:

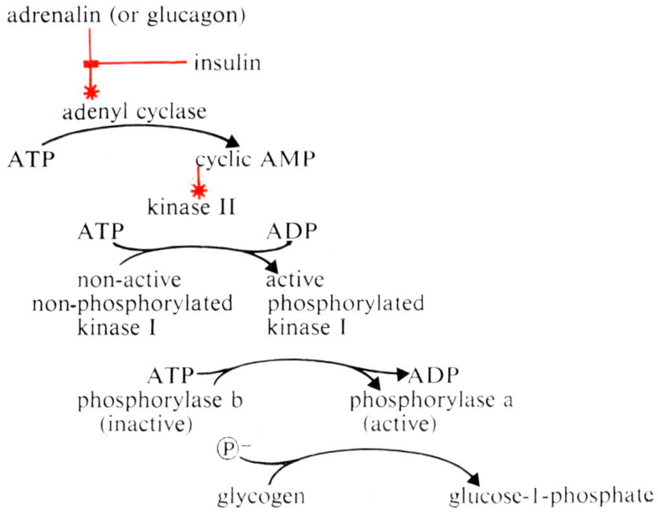

Note that the symbol  is an activation and 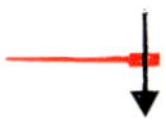 an inhibition.

Thus B 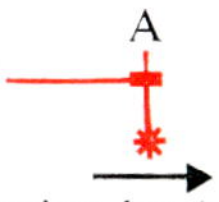

represents the blocking by B of activation by A whereas 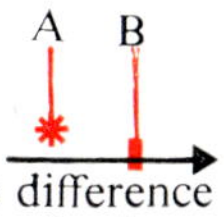

means that A activates while B inhibits. The difference is that in the first case B *alone* has no effect, whereas in the second case it does.

You are not expected to remember this diagram in detail, but you should be able to interpret the response of a system given a diagram of this kind.

The significance of a cascade of this kind is that it provides a kind of *amplification* system. An enzyme that is activated will produce many more molecules of its product than the number of molecules of activator present. This in itself constitutes amplification, but the *gain* of such a system—the ratio of change of concentration of product in the system to change of concentration of activator—is limited. In the case of metabolic control where the activator (or inhibitor) is present in substantial amounts, a single-stage reaction is sufficient. In the case of hormonal control, however, the hormones are present in such tiny amounts that cascades of this sort occur quite frequently.

glycogen synthetase

The control of *glycogen synthetase* is remarkably similar. Again there are two forms of the enzyme, one phosphorylated and the other non-phosphorylated. Again one of these forms is more active than the other under physiological conditions. Again this difference in activity is due to a difference in an allosteric dependence of the enzyme, in this case on glucose-6-phosphate. Again the enzyme is subject to hormonal control, which is exerted upon the kinase reaction which converts the non-phosphorylated enzyme to the phosphorylated one.

ITQ 8 Study the following diagram, which sums up these relationships in the same way as the previous diagram did for phosphorylase.

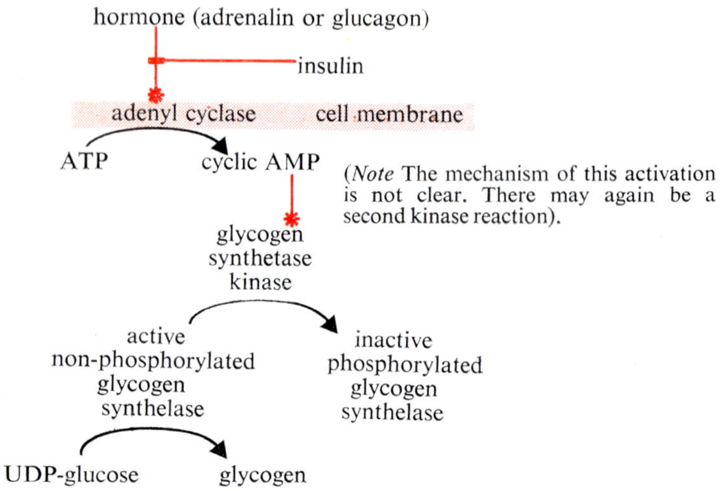

Also refer to the previous diagram on phosphorylase (p. 33) to indicate whether the following statements are true or false:

1 Adrenalin causes the activation of both glycogen synthetase and glycogen phosphorylase.

2 Adrenalin causes the activation of glycogen phosphorylase and the inhibition of glycogen synthetase.

3 Insulin causes the inhibition of a glycogen phosphorylase and the activation of glycogen synthetase.

4 Insulin inhibits the effect of adrenalin on the two enzymes, but has no effect alone.

5 In the presence of adrenalin, insulin will cause a depression of levels of cyclic AMP within the cell.

6 Cyclic AMP is a *second messenger* for both glucagon and adrenalin.

7 Glycogen phosphorylase is under both metabolic and hormonal control, whereas glycogen synthetase is under hormonal control only.

4.8.4 The role of cyclic AMP

This metabolite has been introduced as a second messenger in the control of liver glycogen metabolism by adrenalin, and it was in this system that it was first discovered, by Sutherland and his colleagues in 1957. Up till that time, *in vitro* hormonal effects could only be shown in systems such as liver slices in which the cells were intact.

These workers succeeded in showing that the effect of adrenalin occurs in two stages, each of which can be demonstrated in broken cell preparations. The first stage is the action of the hormone on the membrane fraction of the tissue, to give a heat-stable factor, later identified as cyclic AMP. The effect of cyclic AMP on the enzymes concerned can then be demonstrated in the cell cytoplasm fraction. Continued work on the substance, however, has shown that a great number of hormonal effects upon cells are mediated by cyclic AMP as a second messenger. Examples are lipolysis in adipose tissue, which is stimulated by adrenalin, glucagon, and corticosteroids among others (see Section 4.8.1), the stimulation of the production of the hormones of adrenal cortex by ACTH (adrenocorticotrophic hormone) and the stimulation of the production of thyroxin in thyroid by TSH (thyroid stimulating hormone). In each case, the action of the hormone is on the cell membrane, releasing cyclic AMP within the cell, and the response of broken cell preparations from the tissue to cyclic AMP mimics that of the effects of those hormones that activate adenyl cyclase.

For example, in liver, as has been seen, cyclic AMP initiates glycogenolysis, while, in adipose tissue, lipolysis results.

The converse is also true. For example, in liver, glycogenolysis is initiated both by glucagon and by adrenalin. In muscle, however, the effect of glucagon is very much less in comparison to that of adrenalin. This differential is reflected in the activation of adenyl cyclase of membrane preparations of muscle and liver by the two hormones, glucagon causing much less activation in muscle membranes than in liver membranes.

Cyclic AMP, then, forms a versatile and widespread controlling factor in mammals. It is not confined even to animals or to organisms with hormones, however; it certainly occurs in many micro-organisms, and there is evidence of its involvement in the control of certain enzymes in bacteria.

slime moulds

In another group of organisms, the cellular slime moulds, cyclic AMP seems to play a role more analogous to animal hormones (i.e. as a 'first messenger' rather than as a 'second messenger'). In these organisms, cyclic AMP is involved in the aggregation and adhesion that occurs in the transition from the individual cell stage to the aggregated stage of the life cycle (see Fig. 8). An interesting feature of this behaviour is that the presence of cyclic AMP in the environment itself causes the cells to produce cyclic AMP, making a 'chain reaction' which causes the rapid completion of the transition from the individual to the aggregated stage.

(*a*)

(*b*)

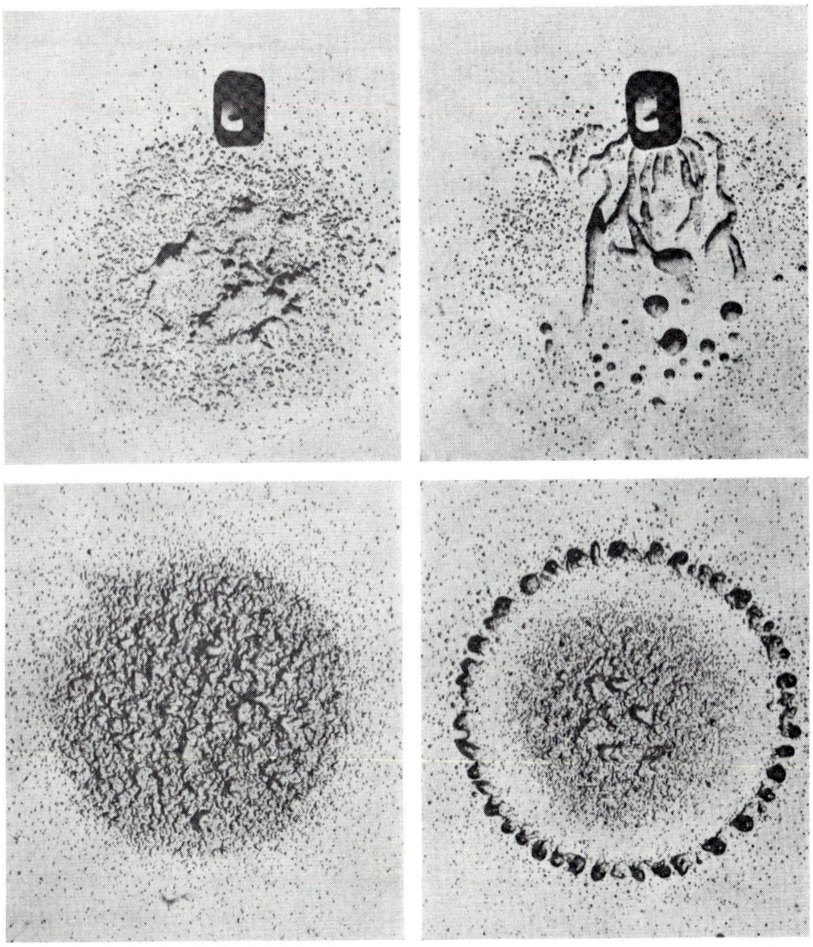

without cyclic AMP *with cyclic AMP*

Figure 8 Cellular slime mould (a) attraction and (b) aggregation stimulated by cyclic AMP

Self-assessment Questions

SAQ 1 (*Objective 2*)
The first stage of glucose catabolism is:
1 Reaction with coenzyme A.
2 Phosphorylation by ATP.
3 The formation of UDP-glucose.

SAQ 2 (*Objective 2*)
The reaction pathway from glucose to lactic acid results in:
1 Production of $NADH_2$ from NAD.
2 Production of $NADPH_2$ from NADP.
3 No production of either $NADH_2$ or $NADPH_2$.

SAQ 3 (*Objective 8*)
Identify the following statements as true or false:
1 Allosteric enzymes have more than one binding site.
2 All regulating enzymes are allosteric enzymes.
3 The enzyme of lowest measured activity in a pathway is a regulating enzyme.

SAQ 4 (*Objective 6*)
Identify the following statements as true or false:
1 Most mammalian cells can convert pyruvic acid to glucose-6-phosphate.
2 Most mammalian cells can convert glucose-6-phosphate to glucose.
3 Liver cells can convert pyruvic acid to glucose-6-phosphate.
4 Liver cells can convert glucose-6-phosphate to glucose.

SAQ 5 (*Objective 2*)
Which of the following represents the stoicheiometry of the pentose phosphate pathway:

1

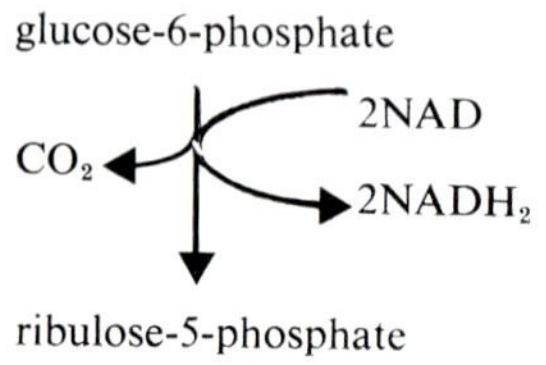

2

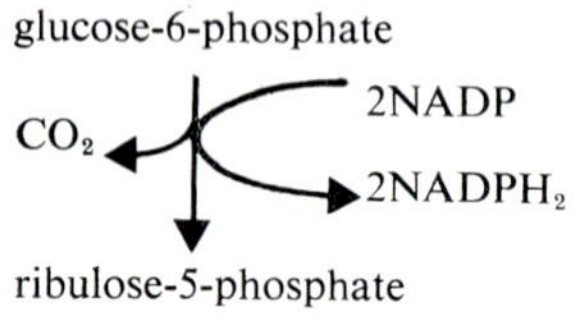

3

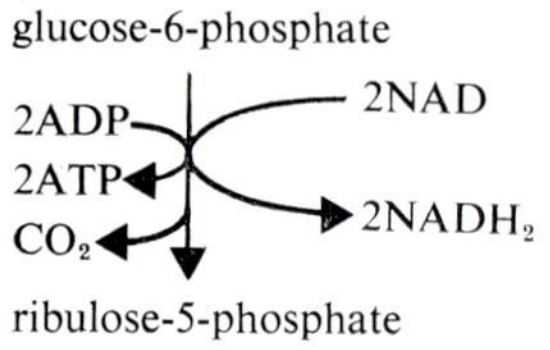

4

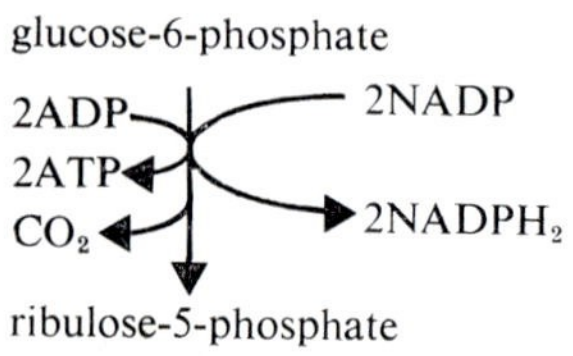

SAQ 6 (*Objective 5*)
This diagram sums up the control points of glucose catabolism:

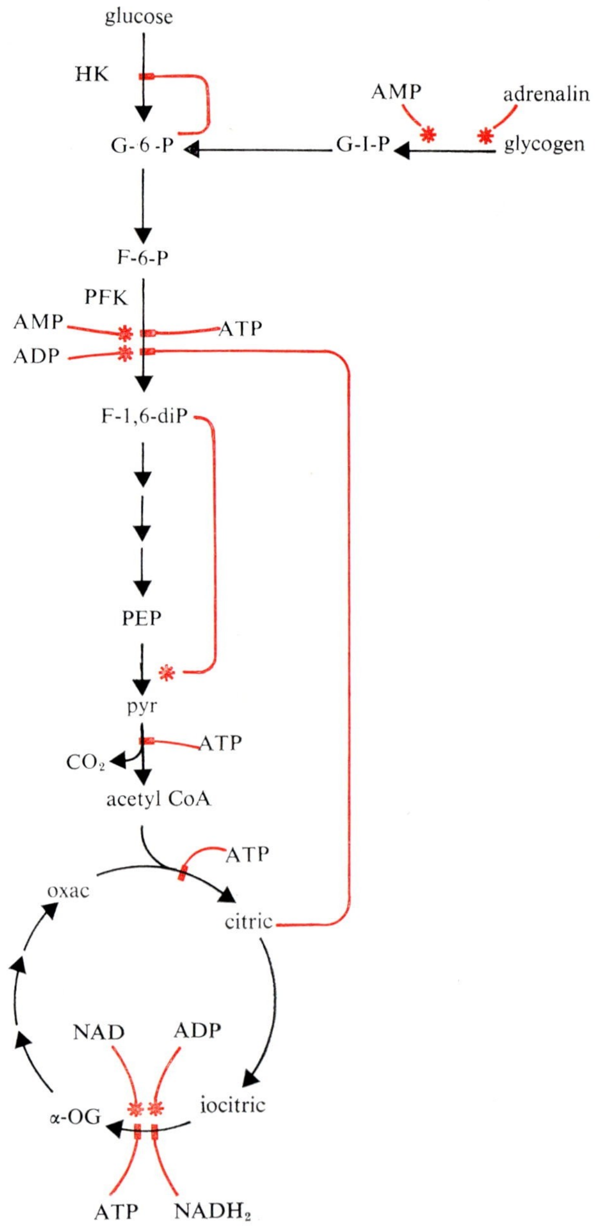

KEY

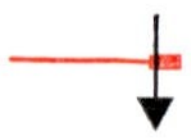 inhibition

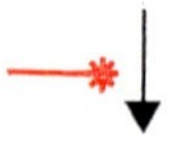 activation

G-1-P	glucose-1-phosphate
G-6-P	glucose-6-phosphate
F-6-P	fructose-6-phosphate
F-1,6-diP	fructose-1,6-diphosphate
PEP	phosphoenolpyruvate
pyr	pyruvic acid
α-OG	α-oxoglutaric acid
oxac	oxaloacetic acid

A number of constraints to the system is listed and, for each, several possible results. For each constraint, label each possible result as (1) expected; (O) not expected; (X) impossible to predict from the data.

1 An *uncoupler* of oxidative phosphorylation is added to the system (refer to Unit 3, Section 3.8, if you have forgotten the action of uncouplers):

(a) Glycolysis will be enhanced.
(b) Glycogenolysis will be inhibited.
(c) The citric acid cycle will be inhibited.
(d) Citric acid will accumulate.
(e) F-6-P levels will decline.
(f) Lactic acid will accumulate.

2 An *inhibitor* of oxidative phosphorylation is added to the system (Unit 3, Section 3.7.3):

(a) Glycolysis will be enhanced.
(b) Glycogenolysis will be inhibited.
(c) The citric acid cycle will be inhibited.
(d) Citric acid will accumulate.
(e) F-6-P levels will decline.
(f) Lactic acid will accumulate.

3 Adrenalin is added to the system:

(a) ATP production will be increased.
(b) ATP production will remain unchanged.
(c) Glucose uptake by the cell will diminish.
(d) Glucose uptake by the cell will remain unchanged.

SAQ 7 (*Objectives 2 and 6*)
Which of the following features apply to fatty acid catabolism and which to fatty acid anabolism?

1 Intermediates are derivatives of coenzyme A.
2 Linked to NAD/$NADH_2$.
3 Proceeds by successive addition or removal of 2-carbon fragments.
4 Intermediates are derivatives of acyl carrier protein.
5 Linked to NADP/$NADPH_2$.
6 Occurs in the mitochondrial fraction.
7 Occurs in the cytoplasmic fraction.
8 Occurs in the ribosomal fraction.

SAQ 8 (*Objective 9*)
Under which of the following conditions would you expect ketosis (or a similar build-up of acetyl CoA or its derivatives) to occur:

1 In a germinating seed, using lipid as its major energy source.
2 In a hypothetical bacterial mutant, lacking pyruvic carboxylase, but possessing the glyoxylic acid sequence, growing on acetyl CoA as sole source of carbon.
3 In a diabetic on a normal diet supplied with a normal therapeutic dose of insulin.
4 In a normal person on a high fat, low carbohydrate and protein diet supplied with excess insulin.

SAQ 9 (*Objective 4*)
In most organisms, under normal conditions, NADP is present mainly as $NADPH_2$ while NAD is present mainly as oxidized NAD. Identify each of the following assertions as (1) supported; (O) contradicted; (X) neither supported nor contradicted by the factual statement:

1 The $NADPH_2$/NADP ratio is not a regulating factor in fatty acid synthesis.
2 The $NADH_2$/NAD is not a regulating factor in mitochondrial oxidative phosphorylation.
3 A possible source of $NADPH_2$ for fatty acid synthesis is the transhydrogenase reaction

$$NADH_2 + NADP \rightarrow NAD + NADPH_2$$

which is freely reversible.
4 A possible source of $NADPH_2$ for fatty acid synthesis is the $NADPH_2$ linked isocitric dehydrogenase reaction

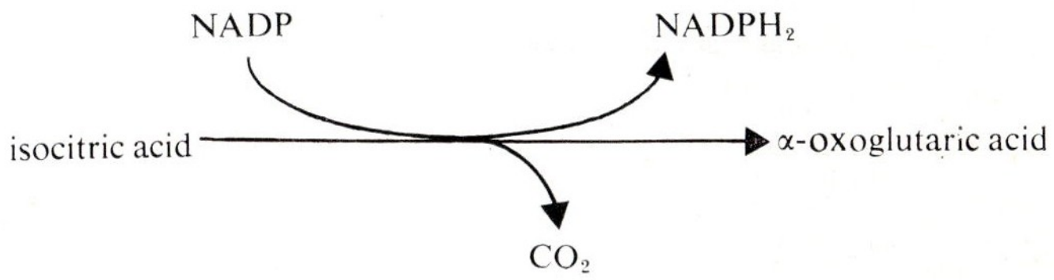

occurring in the cytoplasm after exit of isocitric acid from the mitochondrion.

Answers to Pre-Unit Assessment Test

1

Anabolism	*Catabolism*
Synthesis	Breakdown
Linked to ATP breakdown	Linked to ATP production from ADP and phosphate
Energy-using	Energy-yielding
Linked to NAD production from $NADH_2$	Linked to $NADH_2$ production from NAD

2 (i) (a) pyruvic acid
(ii) (b) ethanol
(iii) (d) lactic acid

3 The citric acid cycle (or Krebs cycle, or tricarboxylic acid cycle).

4 ATP production (from ADP and phosphate)
Reduction of NAD (producing $NADH_2$).

5 Glycolysis—cytoplasm.
Protein synthesis—ribosomes.
Oxidative phosphorylation—mitochondria.
Citric acid cycle—mitochondria.

ITQ Answers and Comments

ITQ 1 (*p. 14*)

(a) 5. There is an equilibration of the trioses which makes carbons 1 and 6 equivalent, and also carbons 3 and 4 equivalent. The 3 and 4 carbons are released first as CO_2 in the step pyruvic acid→acetyl CoA+CO_2.

(b) 1. Carbon 1 of the starting glucose will be lost in the first dehydrogenase reaction, so 1-(^{14}C)-glucose will result in the highest initial rate of $^{14}CO_2$ production. In the sugar transformations that follow, both the resulting hexoses have C–1 derived from C–2 of the original glucose, while the triose has C–1 derived from C–4 of the original glucose. If two triose molecules combine, the resulting hexose will have C–1 derived from C–6 of the original glucose. If only the pentose phosphate pathway is operating, then in the next dehydrogenase reaction two-thirds of the carbon dioxide will result from C–2 of the starting glucose, and one-third from C–6.

ITQ 2 (*p. 19*)

2 and 5 are favoured interpretations:

2 Glycine is a common amino acid, and is joined to each of the compounds excreted by an amide bond. This is shown most clearly in the last example, in which the substance recovered is simply the substance fed combined with glycine.

5 The successive removal of 2-carbon fragments can be represented by

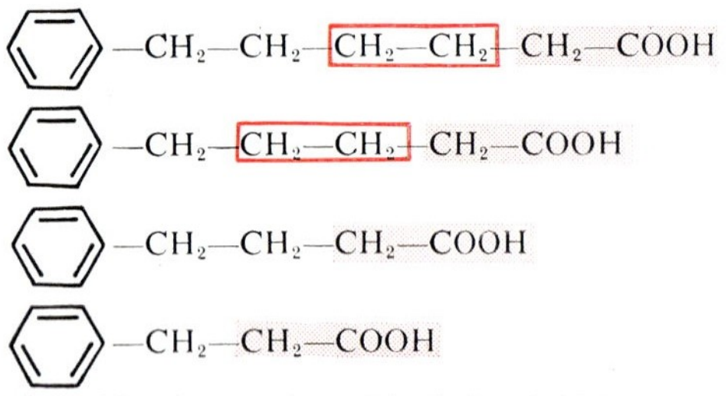

etc., followed by the reaction with glycine (which occurs in the liver).

1 It is not possible to make any comment on this from the data given. In fact, it is untrue, because glycine derivatives or normal fatty acids are not excreted in the urine.

3 This is contradicted by the data.

4 This is conceivable, but much less likely than 5.

6 This is a highly unlikely mechanism, and is contradicted by the evidence from the last substance fed.

The successive removal of two carbon fragments was called by Knoop 'β-oxidation' because he assumed, correctly, that oxidative attack occurred on the β-carbon in the chain:

$$R—CH_2—\overset{\gamma}{C}H_2—\overset{\beta}{C}H_2—\overset{\alpha}{C}H_2—COOH$$

(an obsolete method of labelling the carbon atoms) although this, of course, occurs to the CoA derivative.

ITQ 3 (*p. 26*)

(a) (X). Getting a cell-free system to work can be a long and tedious process.

(b) (O). If the system is working, one would expect malonyl CoA to be used as well as acetyl CoA.

(c) (X). Although the scheme requires acetyl CoA as well it is impossible to say from the data given that acetyl CoA is not being generated from malonyl CoA.

(d) (1). This is in fact what happens.

ITQ 4 (*p. 27*)

(a) and (b) (X). Although it would be expected that label would be incorporated into palmitic acid, it is possible that there might be an interconversion of label acetyl CoA→malonyl CoA which would dilute the label so that it would not be detected, or cause it to be evenly distributed.

(c) (O). This is in direct contradiction to the scheme, which requires the acetyl group carbons to be at the other end of the molecule.

(d) (1). This is in fact what happens (see (c) above).

ITQ 5 (*p. 27*)

(a) (O).

(b) (O).

(c) (1). From the scheme, it is clear that the labelled CO_2 that is added to acetyl CoA to give malonyl CoA is the same carbon as that eliminated when the malonyl group is added to the growing acyl chain.

ITQ 6 (*p. 29*)

1 (X).

2 (X).

3 (1).

While there is a lot of evidence that suggests that insulin has effects *both* on membrane uptake and on intracellular metabolism, the data given here do not allow any conclusions of this sort to be drawn. An intracellular control point can affect uptake of a metabolite just as easily as control of uptake can affect intracellular mechanism.

ITQ 7 (*p. 30*)

Possible interpretations 1

1 (1). The incorporation of glucose increases, and in addition, there is incorporation from acetate. You should bear in mind, however, that incorporation of label is not a direct measure of rate of synthesis (TV4).

2 (O). This is contradicted by the data in the table. Line 1 shows an increase of uptake of glucose carbon in the presence of acetate.

3 (1). Lines 1 and 5 demonstrate this.

4 (O). The sum of lines 2 and 3 are: without acetate 15.95; with acetate 19.75. In both cases, these are significantly less than the uptake into tissue (line 1), 20.85 and 24.6 respectively.

5 (O). The rate of glycolysis is obtained by subtracting the figure in line 4 from that in line 3. This gives a figure for glycolysis of 3.35 without acetate, and 2.30 with acetate. So the rate of glycolysis is *inhibited* by the addition of acetate.

6 (X). The table presents no data on this point.

So the presence of acetate *stimulates* the incorporation of glucose carbon into the tissue, into fatty acid, and somewhat inhibits the rate of glycolysis.

Possible interpretations 2

7 (1). The amount of extra glucose incorporated in the presence of acetate is 24.6—20.85=3.75. The increase in metabolism of glucose by the pentose phosphate pathway in the presence of acetate is 7.7—4.2=3.5. Within experimental error this is the same as 3.75, above. If you got this wrong, check your response to question 9.

8 (1). Since the addition of acetate increases the rate of incorporation into fatty acid, it seems reasonable that the rate of production of acetyl CoA from glucose is rate-limiting.

9 (1). This is more complex. Assume that the principal (or only) way for $NADPH_2$ to be oxidized is *via* fatty acid synthesis.

The pentose phosphate pathway is linked to NADP reduction, i.e. it cannot function without oxidized NADP.

If the supply of oxidized NADP is rate-limiting, factors which increase fatty acid synthesis will tend to increase the rate of metabolism of glucose through the pentose phosphate pathway. Acetate, which increases the rate of fatty acid synthesis, also increases the rate of the pentose phosphate pathway, and so interpretation 9 is supported.

10 (1). The addition of acetate causes an increase in incorporation of label from glucose into fatty acid. You know (Section 4.7.1) that acetyl CoA carboxylase is an allosteric enzyme, stimulated by citric acid. As the level of citric acid could be raised by the addition of acetic acid (via acetyl CoA) it seems likely that the increase of incorporation observed is due to the activation of acetyl CoA carboxylase. If activation of an enzyme is to result in a stimulation of an entire pathway, it must be a rate-limiting step.

11 (O) and 12 (1).

The answers to questions 8, 9 and 10 shows that there must be several different control points in the system.

This exercise is not as clear-cut as the previous ones. Its object is to bring you in touch with the complexities of interpretation of experiments in a field full of controversy, and to prevent you from thinking that biochemistry is a neat subject.

ITQ 8 (*p. 34*)

1 False
2 True
} If you got these wrong, check that although in both cases the adrenalin causes a phosphorylation of the enzyme, in the case of phosphorylase the phosphorylated form is the *active* one, in the case of the synthetase the phosphorylated form is inactive.

3 False
4 True
5 True
} Check the meaning of the symbols used if you got this wrong. Insulin *opposes the activation* of adenyl cyclase by adrenalin, but does not itself inhibit the enzyme.

6 True
7 False. The two forms of synthetase are dependent on glucose-6-phosphate in a way analogous to the dependence of phosphorylase on AMP. However, it is still not clear whether changes in glucose-6-phosphate are sufficient to cause any substantial changes in the activity of this enzyme *in vivo*.

Self-assessment Answers and Comments

SAQ 1
2 The reaction catalysed by hexokinase.

SAQ 2
3 The $NADH_2$ produced in the first stage of glycolysis is oxidized in the reaction from pyruvic acid to lactic acid.

SAQ 3
1 True (see Unit 5 for a more detailed treatment of this).
2 False. Regulating enzymes can also be under hormonal control.
3 False. The enzyme of lowest activity may have been only partially extracted or partially inactivated, or not assayed under ideal conditions.

SAQ 4
1 True.
2 False. Only liver, intestine and kidney cortex can do this.
3 True.
4 True.

SAQ 5
2 The reaction is NADP-linked, and does not produce any ATP.

SAQ 6
1 The uncoupler prevents phosphorylation of ADP to ATP. So $NADH_2$ oxidation will proceed normally, or at an accelerated rate, but ATP will not be produced so the ATP/ADP ratio will fall, causing activation of PFK and phosphorylase. Glycolysis and glycogenolysis will thus increase, and F-6-P levels will decline.

The control of the isocitric dehydrogenase is by both ATP/ADP and $NADH_2$/NAD ratios. In the presence of the uncoupler, the first falls, the second either falls or stays constant. Isocitric dehydrogenase is activated in either case, so the citric acid cycle is enhanced. The $NADH_2$ produced by glycolysis and the citric acid cycle, is oxidized normally under these conditions, so no lactic acid will accumulate. Answers are:

(a) 1	(d) 0
(b) 0	(e) 1
(c) 0	(f) 0

2 Under these circumstances, neither $NADH_2$ oxidation nor ATP production by oxidative phosphorylation can occur. The $NADH_2$/NAD ratio will *rise* and the ATP/ADP level fall. PFK and glycogen phosphorylase will be activated as before, increasing the rates of glycolysis and glycogenolysis, and causing a decline in F-6-P levels. Now, however, the citric acid cycle will be inhibited by the absence of NAD (a *mass action* effect). The $NADH_2$/NAD and ATP/ADP ratios are now acting in opposite directions on isocitric dehydrogenase and so the effect on citric acid accumulation cannot be predicted. The accumulation of $NADH_2$ will result in reduction of pyruvic acid to lactic acid. In effect, the cell becomes anaerobic.
Answers are:

(a) 1	(d) X
(b) 0	(e) 1
(c) 1	(f) 1

3 The primary effect of adrenalin is to activate glycogen phosphorylase, which breaks glycogen down to G-1-P, and hence to G-6-P. The increased level of G-6-P will inhibit hexokinase (HK). Whether the *overall* rate of G-6-P metabolism (glycogen plus glucose) will increase significantly cannot be judged from the data.
Answers are:

(a) X	(c) 1
(b) X	(d) 0

SAQ 7

1 Catabolism.	5 Anabolism.
2 Catabolism.	6 Catabolism.
3 Anabolism and catabolism.	7 Anabolism.
4 Anabolism.	8 Neither.

SAQ 8
Only in (4).
Both seed and the bacteria have the glyoxalic acid sequence, which enables them to use lipid or acetic acid as sole carbon source. Pyruvic carboxylase is not needed as an anaplerotic route provided the glyoxylic acid sequence is present.

A diabetic on the normal therapeutic dose of insulin metabolizes carbohydrate normally. Insulin, however, can only aid the metabolism of carbohydrate that is present in adequate quantities to maintain the citric acid cycle—it cannot itself make up for a carbohydrate deficiency.

SAQ 9
1 (1). If the $NADPH_2$ is fully reduced, its level cannot fluctuate enough to act as a controlling factor, as *small* changes in amount represent a *small* relative change.
2 (X). However, it is quite possible for $NADH_2$ to be a controlling factor, as *small* changes in amount of reduced $NADH_2$ represent a *large* relative change. The data, however, do not provide information as to whether this does occur.
3 (O). The data contradict this, as if the reaction is readily reversible the high $NADPH_2$/NADP combined with low $NADH_2$/NAD will push the reaction the other way.
4 (X). This is a possible reaction, and has been postulated as a significant source of $NADPH_2$. Again, however the data here are irrelevant.

References to S100

1 Unit 15, *Cell Dynamics, and the Control of Cellular Activity*, Section 15.5.
2 Unit 15, Section 15.6.3.
3 Unit 18, *Cells and Organisms*, Section 18.3.1.
4 Unit 15, Section 15.7.3.
5 Unit 10, *Covalent Compounds*, Section 10.2.
6 Unit 18, Section 18.4.

Acknowledgement

Grateful acknowledgement is made for the following illustration used in this Unit:

Fig. 8 University of Chicago Press for J. M. Konijin, *The American Naturalist*, **102**, 1968.

Biochemistry

1 Biological Macromolecules

2 Enzymes

3 Cell Energetics; the Mitochondrion and the Chloroplast

4 Metabolic Pathways

5, 6 } Regulation of Cell Processes

OVERLAY B

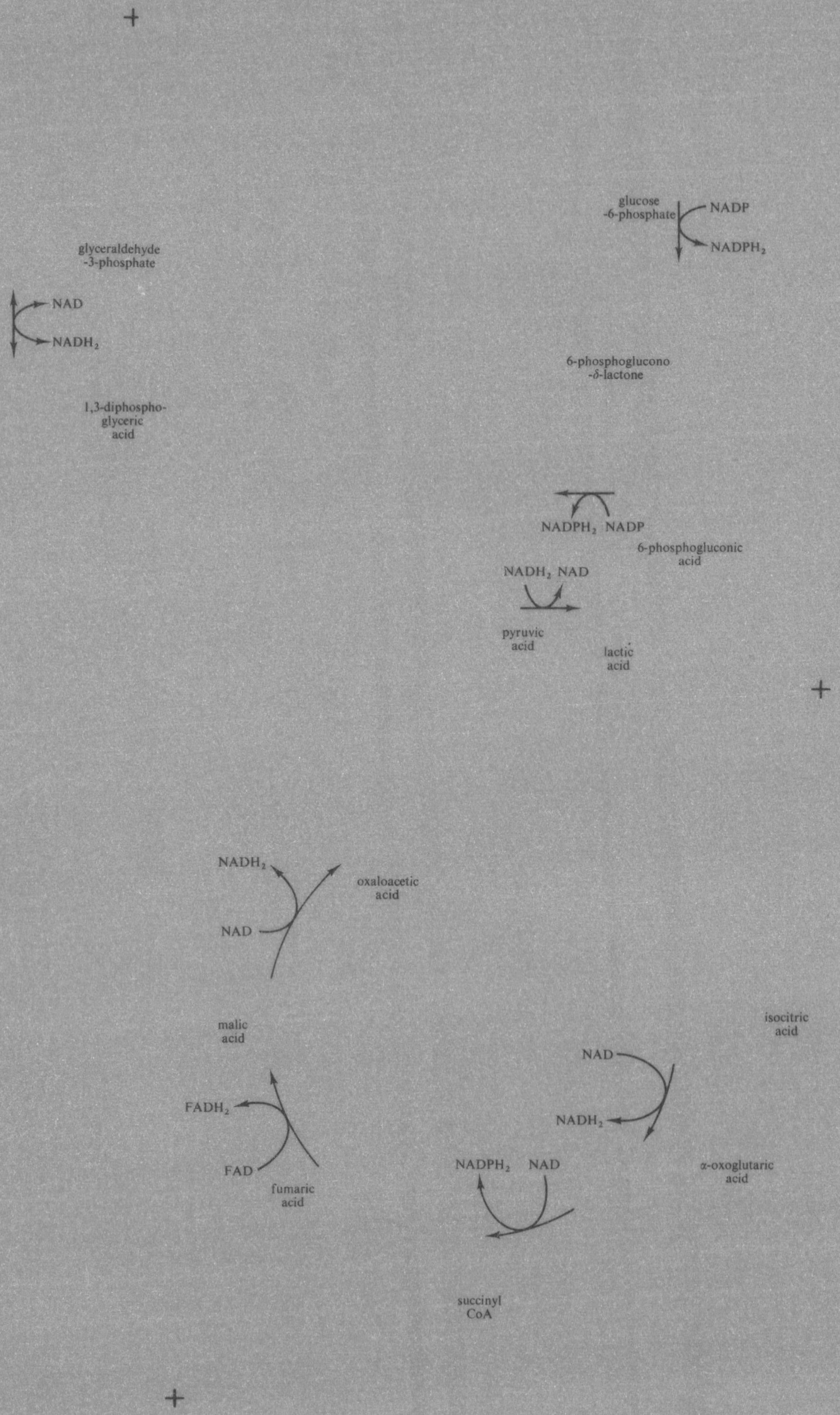

glucose -6-phosphate
NADP
NADPH$_2$
glyceraldehyde -3-phosphate
NAD
NADH$_2$
6-phosphoglucono -δ-lactone
1,3-diphospho- glyceric acid
NADPH$_2$ NADP
6-phosphogluconic acid
NADH$_2$ NAD
pyruvic acid
lactic acid
NADH$_2$
oxaloacetic acid
NAD
malic acid
isocitric acid
NAD
FADH$_2$
NADH$_2$
FAD
fumaric acid
NADPH$_2$
NAD
α-oxoglutaric acid
succinyl CoA

OVERLAY A

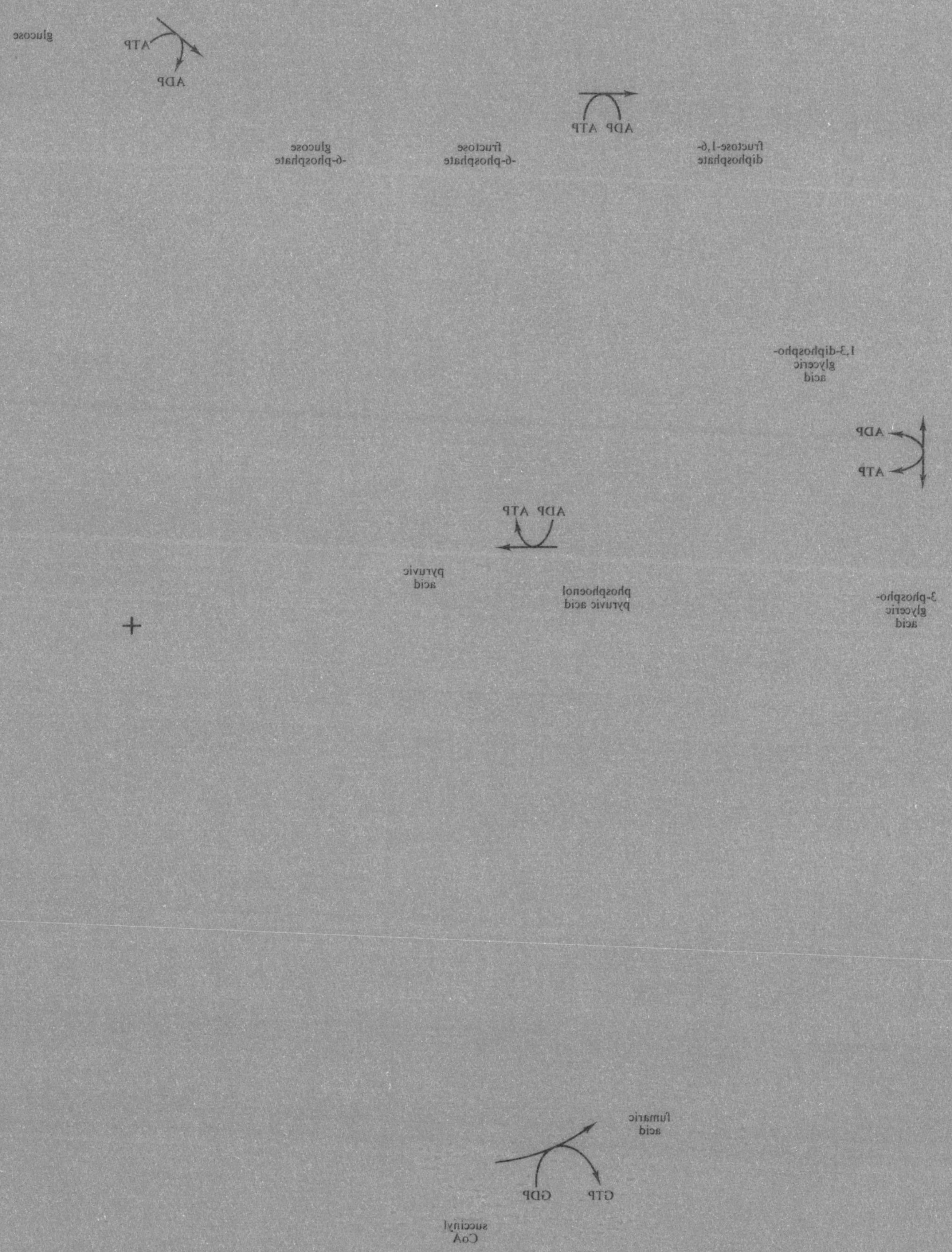

glucose
ATP
ADP
glucose
-6-phosphate
fructose
-6-phosphate
ATP
ADP
fructose-1,6-
diphosphate
1,3-diphospho-
glyceric
acid
ADP
ATP
3-phospho-
glyceric
acid
phosphoenol
pyruvic acid
ADP
ATP
pyruvic
acid
succinyl
CoA
GDP
GTP
fumaric
acid

OVERLAY A

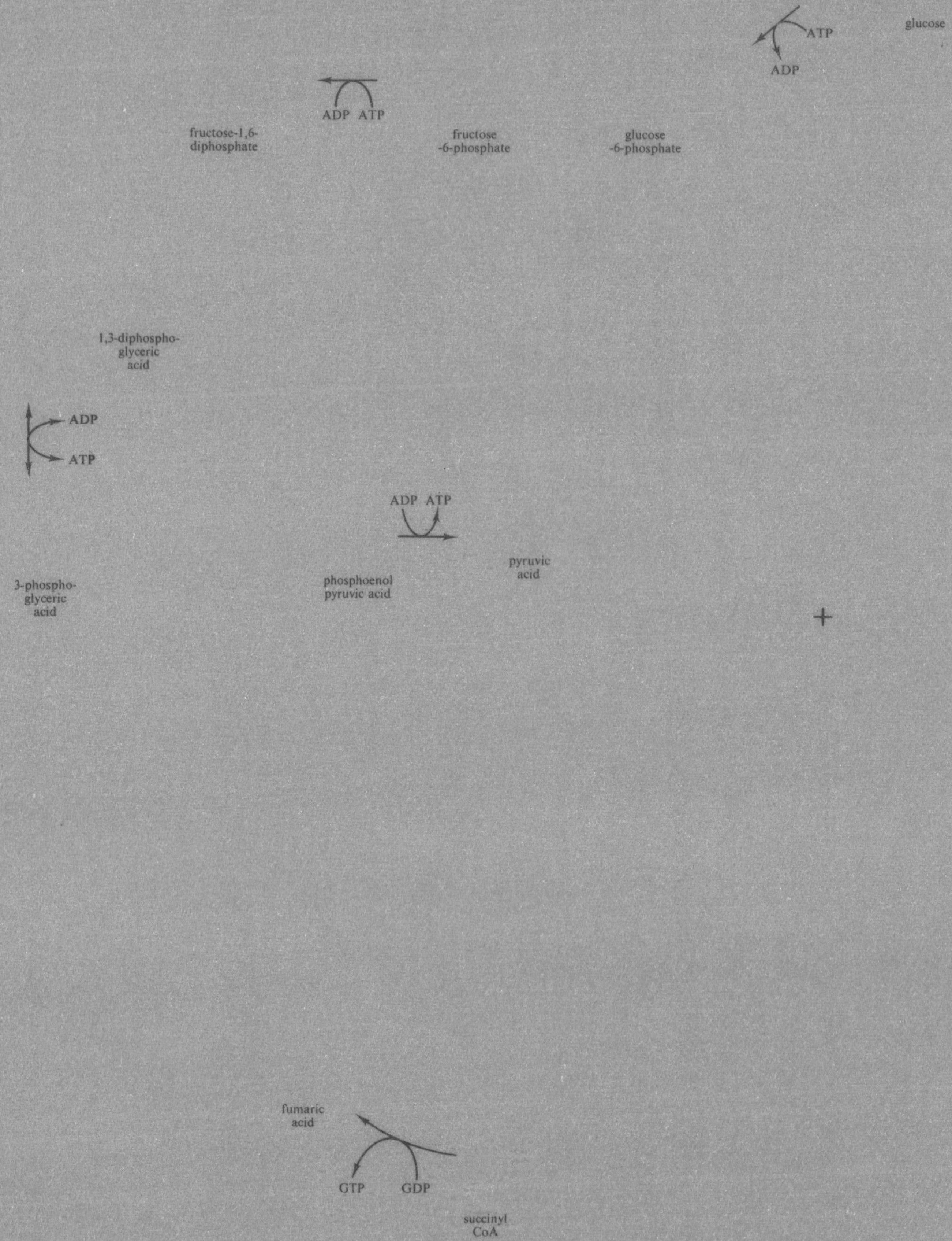
glucose
ATP
ADP
ADP ATP
fructose-1,6-
diphosphate
fructose
-6-phosphate
glucose
-6-phosphate
1,3-diphospho-
glyceric
acid
ADP
ATP
ADP ATP
pyruvic
acid
3-phospho-
glyceric
acid
phosphoenol
pyruvic acid
fumaric
acid
GTP
GDP
succinyl
CoA

OVERLAY C

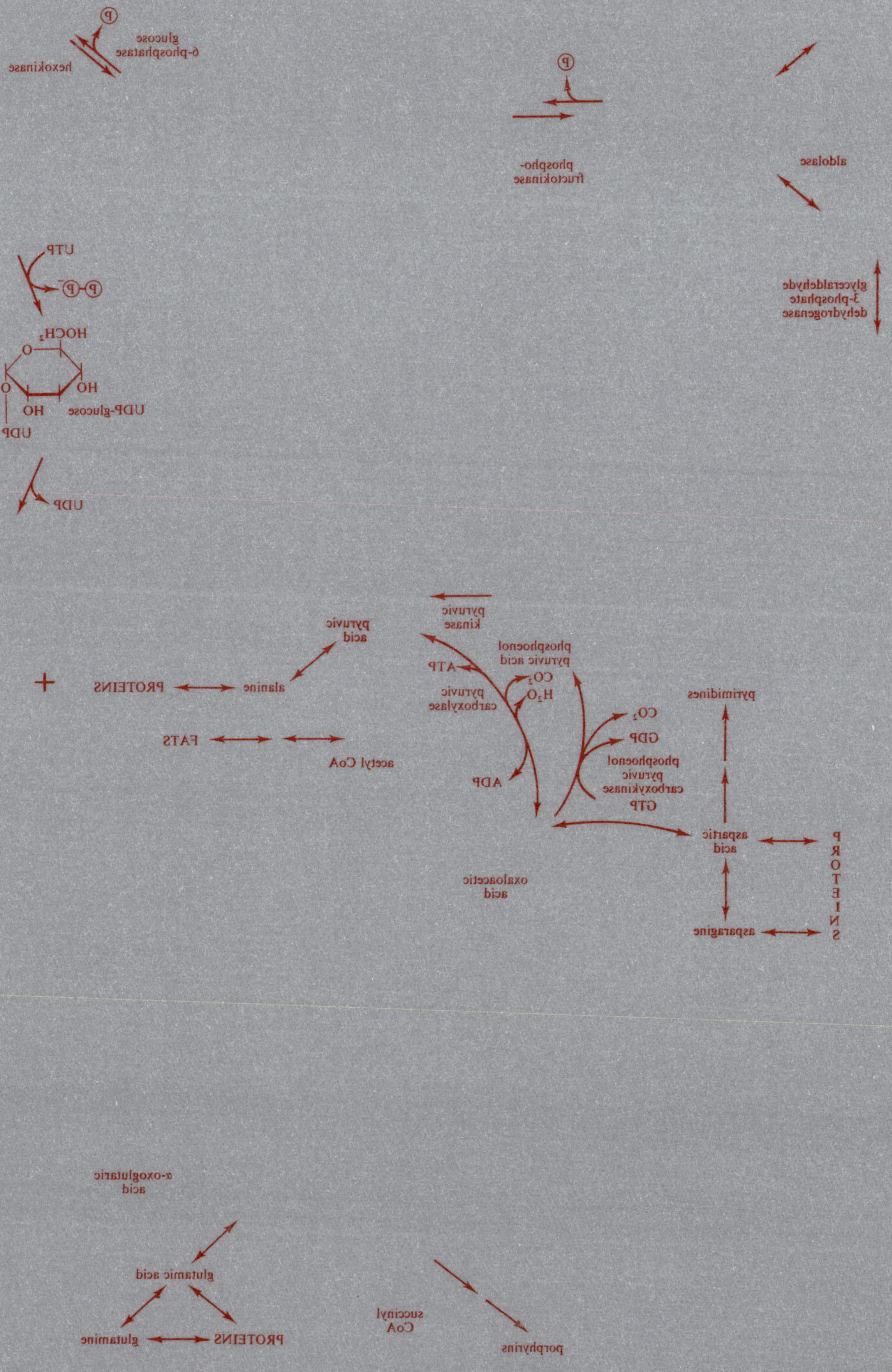
hexokinase
glucose
6-phosphatase
phospho-
fructokinase
aldolase
glyceraldehyde
3-phosphate
dehydrogenase
UTP
HOCH2
UDP-glucose
UDP
pyruvic
kinase
pyruvic
acid
phosphoenol
pyruvic acid
ATP
CO2
H2O
pyruvic
carboxylase
ADP
alanine
PROTEINS
FATS
acetyl CoA
CO2
GDP
phosphoenol
pyruvic
carboxykinase
GTP
pyrimidines
aspartic
acid
asparagine
PROTEINS
oxaloacetic
acid
α-oxoglutaric
acid
glutamic acid
glutamine
PROTEINS
succinyl
CoA
porphyrins

OVERLAY C

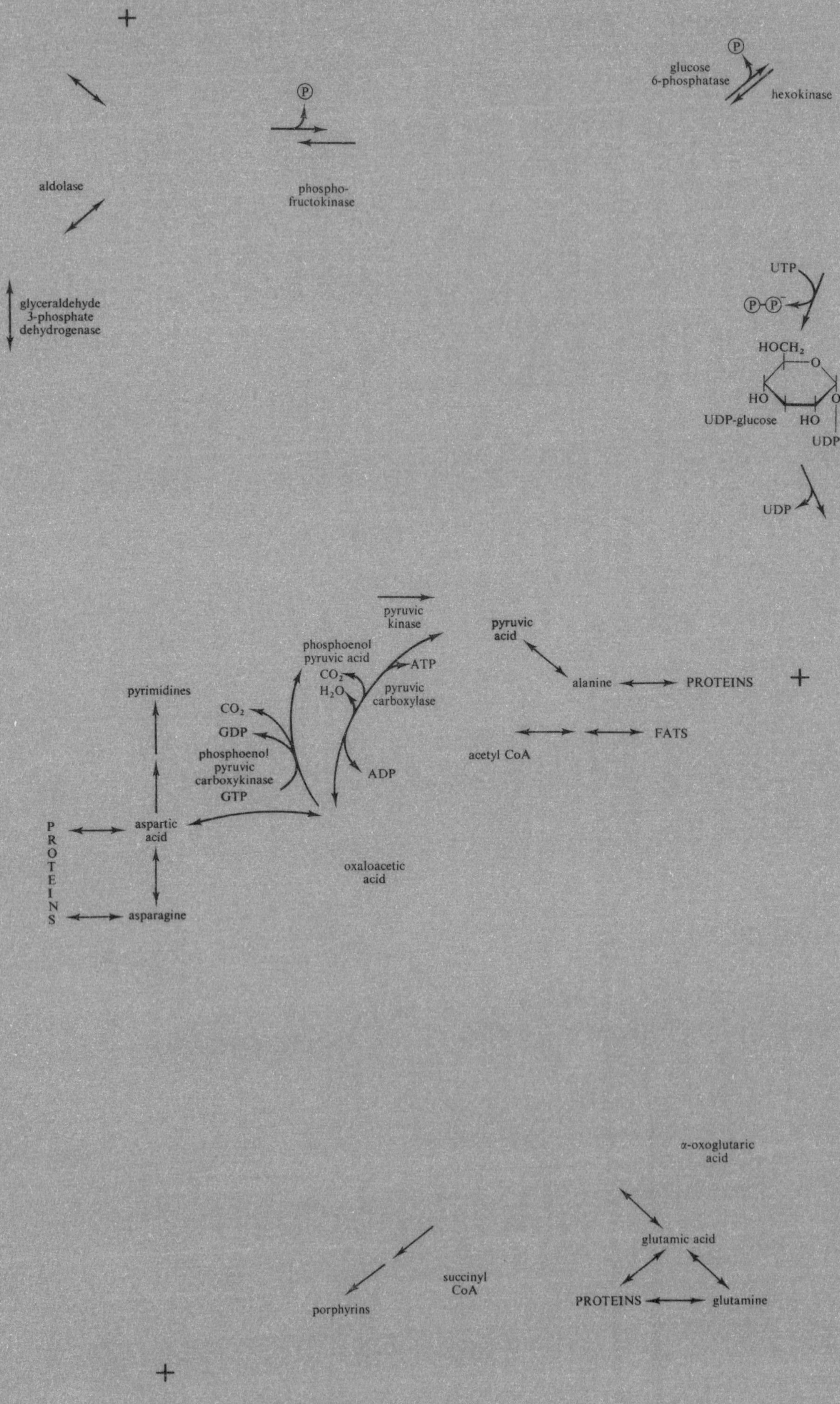

glucose
6-phosphatase
hexokinase
aldolase
phospho-
fructokinase
UTP
HOCH$_2$
HO
O
UDP-glucose
HO
UDP
UDP
glyceraldehyde
3-phosphate
dehydrogenase
pyruvic
kinase
pyruvic
acid
phosphoenol
pyruvic acid
ATP
CO_2
H_2O
pyruvic
carboxylase
alanine
PROTEINS
pyrimidines
CO_2
GDP
phosphoenol
pyruvic
carboxykinase
GTP
ADP
acetyl CoA
FATS
PROTEINS
aspartic
acid
oxaloacetic
acid
asparagine
α-oxoglutaric
acid
glutamic acid
succinyl
CoA
porphyrins
PROTEINS
glutamine

OVERLAY B

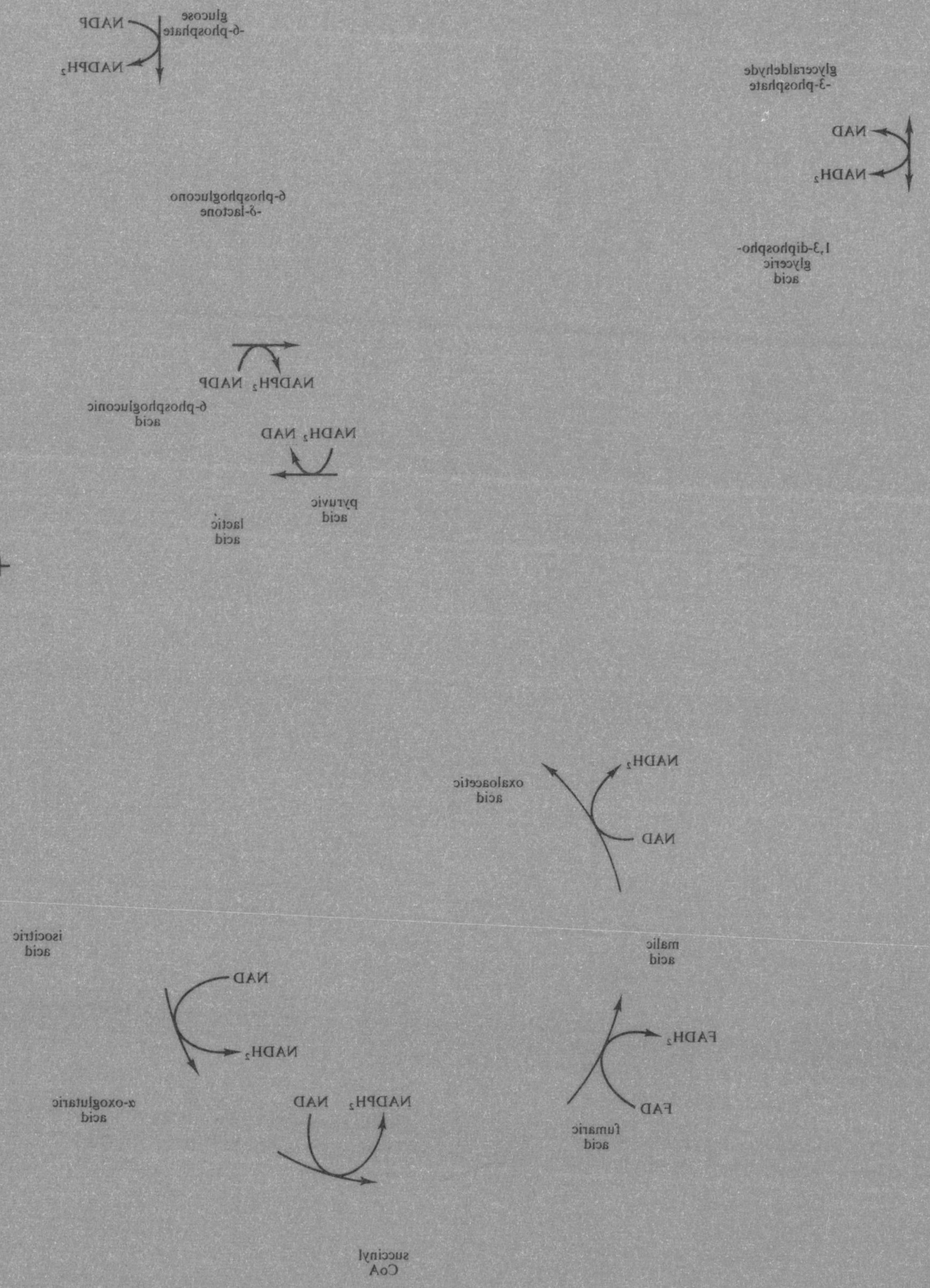
glucose -6-phosphate
NADP
NADPH$_2$
6-phosphoglucono -δ-lactone
NADPH$_2$ NADP
6-phosphogluconic acid
glyceraldehyde -3-phosphate
NAD
NADH$_2$
1,3-diphospho- glyceric acid
NADH$_2$ NAD
pyruvic acid
lactic acid
oxaloacetic acid
NADH$_2$
NAD
malic acid
FADH$_2$
FAD
fumaric acid
isocitric acid
NAD
NADH$_2$
α-oxoglutaric acid
NADPH$_2$ NAD
succinyl CoA